Irrationally Yours

ALSO BY DAN ARIELY

Predictably Irrational
The Upside of Irrationality
The Honest Truth About Dishonesty

Irrationally Yours

On Missing Socks, Pickup Lines, and Other Existential Puzzles

Dan Ariely

WITH CARTOONS BY
WILLIAM HAEFELI

(All profits from this book support research)

HARPER PERENNIAL

NEW YORK • LONDON • TORONTO • SYDNEY • NEW DELHI • AUCKLAND

HarperCollins books may be purchased for educational, business, or sales promotional use. For information please e-mail the Special Markets Department at SPsales@harpercollins.com.

FIRST EDITION

Designed by Kris Tobiassen of Matchbook Digital

Library of Congress Cataloging-in-Publication Data

Ariely, Dan.
   Irrationally yours : on missing socks, pickup lines, and other existential puzzles
/ Dan Ariely ; with cartoons by William Haefeli.—First edition.
      pages cm
   ISBN 978-0-06-237999-3 (trade pbk.)—ISBN 978-0-06-238001-2 (ebook)
   1. Reasoning (Psychology)—Miscellanea. 2. Logic—Miscellanea. 3. Irrationalism
(Philosophy)—Miscellanea. I. Title.

   BF442.A75 2015
   153.4—dc23   2014045382

15 16 17 18 19   OV/RRD   10 9 8 7 6 5 4 3 2 1

To the oddities, complexities, and beauty of human nature

Contents

New Questions and Advice

Introduction

Here is a possible rationalization: My ability to observe and reflect on human nature is rooted in my injury and its continuing effects—thanks to being ripped out of my teenage life, sustaining third degree burns on about 70 percent of my body, being hospitalized for almost three years, experiencing substantial daily pain, experiencing over and over the dysfunction of the medical system, and having extensive scars that make me feel out of place in most social circumstances. Combined, these elements (so the rationalization goes) have made me a better observer of life. It is also what brought me to study social science.

Don't get me wrong; I don't think that my injury was worth it. No one can rationalize this much pain and misery. But the complex experiences of my injury, time in the hospital, and life with extensive scars and disabilities have been my microscope on life. Through this perspective I have been able to observe substantial human suffering. I have seen people who managed their suffering and triumphed, and I have seen those who caved in. I have been exposed to different medical procedures and odd human interactions. From the distance of my hospital bed, I was able to observe the people around me go about their normal life, to wonder about human habits, and to speculate about the reasons we act the way we do.

Because of my scars, the pain, some odd-looking medical braces, and the pressure bandages that covered me from head to toe, the feeling of living separately from the normal day-to-day life did not stop when I left the hospital. As I made my first steps back into the reality that I once took for granted, my viewpoint expanded to include more mundane daily activities such as the way we shop, drive, volunteer, interact with coworkers, take risks, fight, and behave thoughtlessly. And, of course, I could not help but notice the intricate fiber that governs romantic life.

With this perspective, I turned to study psychology. Very soon my personal and professional life became deeply intertwined. I remembered placebo medications for pain, and I conducted experiments to better understand the effects of expectations on painful treatments. I remembered some of the bad news I got while in the hospital, and I tried to figure out how best to break bad news to patients. There were many other topics that crossed the personal/professional boundary, and over time I learned more and more about my own decisions and the behaviors of those around me. This was more than twenty-five years ago and since then I have dedicated most of my time to try to better understand human nature, focusing mostly on where we make mistakes and what can be done to improve our decisions, actions, and outcomes.

After writing academic papers on these topics for many years, I started writing about my research and its implications in a more conversational and less academic way. Perhaps because I described how my research grew out of my own difficult experiences, many people started sharing with me their personal struggles. Sometimes they were curious to know what social science can say about a particular experience they had, but most often these were questions about their own challenges and decisions.

While responding to as many of these requests as I could, it became apparent to me that some of these questions were of general interest. And in 2012, with permission of the people asking the questions, I started answering some of the broader questions publicly through my *Wall Street Journal* column "Ask Ariely." The book you are now holding in your hands includes some edited and expanded answers from the column in addition to some questions and answers that have not appeared previously in press. Most important, this book also includes some wonderful cartoons by the talented William Haefeli that, in my opinion, deepen, improve, and expand my answers.

Now you have it. Aside from my ability to rationalize, does any of this make my advice more valuable, accurate, or useful? I'll let you be the judge.

Irrationally yours,
DAN ARIELY

Irrationally Yours

ON ESCALATION
OF COMMITMENTS

Dear Dan,

Every year, when Christmas comes, I feel obligated to send Christmas cards to everyone I know, and every year, the number of cards I send gets larger and larger. It is now officially getting out of hand. Can I switch to sending cards only to my really close friends?

—HOLLY

A few years ago I was ordained by the Church of Spiritual Humanism, so I feel that I am in the position to tell you that it is perfectly fine for you to send cards only to your good friends. As a social scientist, I don't think anyone left off the list will be offended and many of them might not even notice. Taking this step will also reduce their feeling of obligation to send you a card next year—so in the process you are also helping them with the same problem. And if you really want to eliminate the Christmas-card frenzy, there is always Judaism.

Friends, Expectations, Happiness

ON THE ART AND JOY
OF SAYING NO

"And once they'd finished everything on their to-do lists,
they lived happily ever after."

Dear Dan,

I've recently been promoted, and I now receive all sorts of requests for activities that have little to do with my love for my job. I recognize the importance of helping out coworkers and the organization as a whole, but these other activities are taking up too much of my time and making it impossible for me to do my job. How can I set my priorities better?

—FRANCESCA

Ah yes—the perils of success. Promotions usually sound good, but once we get them, we often realize that they come with extra demands and annoyances. (Oddly, we don't seem to remember this lesson from promotion to promotion, so each time we're surprised when we discover these extra costs.)

Back to your question. Here's what I suspect your new life looks like: Every day, nice coworkers who you want to help ask you to do something for them. On top of that, the request is usually for some time that is far into the future—say a month from now. You look at your calendar and it looks rather empty, so you say to yourself, "Since I'm mostly free a month from now, how can I say no?" But this is wrong. Your future is not really going to be free; the details are just not yet filled in. When the day arrives, you will have all kinds of things to do and your plate will be overflowing—even without the extra demands of this request. And at that point you will wish that you hadn't said yes.

This is a very common problem, and I would like to propose three simple tools that can help you better stick to your desired priorities.

First, every time a request comes in, ask yourself what you would do if it was for next week. Framed this way, you would look at your schedule and figure out if you would cancel some of your other obligations to make room for this new request. If you would cancel some things in order to make time, go ahead and accept. But, if you would not prioritize it over your other obligations, just say no.

As a second tool: When you receive a request, imagine that you look at your calendar to see if you can comply and you discover that you are fully booked that day without the ability to switch anything around—maybe you are out of town. Now, try to gauge your emotional reaction to this news. If you feel sad, you should go ahead and accept the request. On the other hand, if you feel relieved that you can't do it, turn it down.

Finally, practice using one of the most beautiful words in English: "cancel-elation," the glee we feel when something is canceled. To use this tool, imagine that you accepted this particular request but it later got canceled. If you can taste joy, you just experienced cancel-elation and you have your answer.

Workplace, Decisions, Long-Term Thinking

ON NETFLIX DISSATISFACTION

"If you had ten eggs in a basket and two fell out and broke,
how sad would you be?"

Dear Dan,

I am a longtime Netflix user. Recently, Netflix removed about 1,800 movies from its offering, while adding a few very good ones. I know I probably would have never watched any of those 1,800 movies but I am still upset about this and I am seriously considering leaving Netflix. Why do I feel this way?

—KRISTEN

As a movie fan myself, I appreciate your conundrum. The basic principle behind this emotional reaction to the elimination of these movies is loss aversion. Loss aversion is one of the most basic and well-understood principles in social science. The basic finding is that losing something has a stronger emotional impact than gaining something of the same value. Going back to Netflix, the implication is that having movies taken away from your account is perceived as a loss and because of that, it feels much more painful. The impact of loss aversion could be so strong that losing the not-so-great movies can still be more upsetting than the joy of getting movies that are objectively better.

One other implication of loss aversion is that while old Netflix users, such as yourself, will view the new collection of movies on Netflix in a somewhat negative and loss-aversive way, new users who just see the new set of movies without the experience of having anything taken away from them will view the updated offering in a much more positive way.

With this in mind, my suggestion is that you try to think about Netflix more like a museum. As a service that provides you not with particular movies but with an optimal, curated variety of entertainment. With museums, we don't think about ourselves as owning any of the art, so we aren't upset when the exhibits change. My guess is that if you manage to reframe your perspective this way, you will enjoy Netflix to a larger degree.

Entertainment, Loss Aversion, Value

ON DIETING

"I'm not eating. I'm self-medicating."

Dear Dan,

This is probably a very common question, as everyone seems to be on a diet at some point. The question is: Why do we let the immediate pleasure of eating overwhelm our long-term considerations? Why do we sabotage our health over and over? And how can we tame our desire to eat and overeat?

—DAFNA

As you pointed out, dieting basically goes against our inherent nature. It's often the case that we have fantastic ideals about our future selves. What we will do, what we will not do, what decisions we will make, and what decisions we will not make. But when it comes to our everyday decisions, often the short-term considerations prevail and our long-term hopes and wishes take a backseat (sometimes they even move to the trunk of the car). When we're not hungry and someone asks us how many desserts we will eat over the next month, we might think that we will have one or maybe two max. However, when we're at a restaurant and the waiter places the dessert menu (or even worse, the dessert tray) in front of us and we see our favorite dessert listed as an option, we get a very different idea about the importance of having dessert right now. We see the triple chocolate cake, and our priorities change. In behavioral economics, we call this "present-focus bias."

On top of that, a diet is really a difficult thing to stick to—much more difficult than quitting smoking, for example. Why? Because with smoking, we are either smokers or nonsmokers. With dieting, we can't easily separate ourselves into eaters and noneaters. We have to eat, and so the question becomes: What do we eat and when exactly do we stop? And because there are no clear-cut stopping rules, it becomes very hard to stick to any particular diet.

So what can we do about this problem? The simplest approach is to realize the extent of this challenge, and try hard from the get-go to avoid exposing ourselves to the types of foods that are detrimental to our diet. If we have no cake at home, we'll probably eat much less cake. And if we replace that cake with fresh bell peppers, we'll eat peppers because they're available. Maybe we can decide that dessert is just unacceptable. Or maybe that we can have dessert only on the Sabbath. Another useful and relatively simple rule is not to let any soft drinks and boxed snacks into our homes. Such an approach, of applying strict, religious-type rules to dieting, can be very useful. By adopting such rules we will be able to better recognize at any point if we are sticking to our long-term plans or not, and this should help us reinforce our desired behavior.

Dieting, Self-Control, Rules

ON FORGOTTEN
AND FORGIVEN LOANS

Dear Dan,

Many years ago a friend of mine asked me to lend her a substantial amount of money. At the time I was happy to help her, but it has been years since I lent her the money, she has never mentioned it, and the shadow of this exchange is clouding our relationship. What should I do? Should I say something?

—MARIEL

Because you're the one who did her a favor and loaned her the money, you probably think that she is the one with the obligation to bring up the topic. This might be true from a moral perspective, but the problem is that once you lent her the money, you shifted the power in your relationship, and this asymmetry is making it much, much harder for her to bring up the topic.

Someone unquestionably should bring this up, and given the asymmetrical power dynamics, I think it should be you.

Now that we have decided that you should bring it up, the next question is what to say. If you need the money, I would say something like, "A few years ago I was happy to loan you some money, but I'm trying to sort out my accounts in the next few weeks and I just need to know when would be a good time for you to repay me." If you don't need the money and are willing to give it to your friend, I would say something like, "A few years ago, you asked me for some money, and I just wanted to make sure that it was clear that I always meant it as a gift."

 Either way, bringing the topic up will be a bit uncomfortable in the short run, but in the long run, it could save your friendship.

Money, Friends, Giving

ON MARRIAGE
AND ECONOMIC MODELS

"It wouldn't be fair for us to get married until gays can be legally married, too. That's my story and I'm sticking to it."

Dear Dan,

An economist friend once said to me that marriage is like betting someone half of everything you own that you'd love your partner forever. Do you agree?

—SHANE

Economists have a lot of interesting ways to look at human behavior. Some of these are deeply misguided but even when these models are wrong, they are often interesting and sometimes useful. Framing marriage as a gamble is a great example of an economic perspective that is both misguided and useful. Describing a deep social and romantic bond as a bet one would place at a casino table leaves much of human relationship by the wayside (the misguided part), but it also emphasizes the large potential for loss, which people often don't take into account when they decide to get married (the useful part). Overall, I suspect that this particular perspective on marriage is more wrong than helpful, but here are three things I am sure about:

First, while describing marriage as a bet could be a useful tale of caution, it is not the way married people think about their joint life, their kids, their commitment, and their future plans. Second, I am also sure that while it might be fun and interesting to think about other people's marriages in terms of bets, we shouldn't be tempted to think about our own relationships in this way. And finally, I am also sure that we shouldn't even mention this perspective to our significant other.

Relationships, Expectations, Predictions

ON SOCIAL NETWORKS
AND SOCIAL NORMS

"Lazy? I've been social-networking my ass off."

Dear Dan,

What is the function of the "Like" button on Facebook? Why doesn't Facebook have options for "Dislike" or "Hate," for example?

—HENRY

Facebook's "Like" button is much more than a way for us to react to other people. It is a social-coordination mechanism that tells us how we can, and should respond. It subtly gives us instructions on what is OK (and not OK) to post and it gently tells us how we can and can't behave on Facebook. Adding buttons such as "Dislike" or "Hate" would change our mindset when we read different posts; it would prompt us to have more negative reactions and I suspect that very quickly it would destroy this social network's positive atmosphere. And for what it's worth, my preference would be to add a button for "Love."

....

Dear Dan,

I graduated from college a few years ago. Since then my social life has been limited to Facebook. And it is far from satisfying.

—JAMES

Facebook has many wonderful aspects but I agree with you that it is no substitute for face-to-face human contact.

While you were in college you probably had a vibrant social life, but you probably also accumulated student loans. Now the social part is over, and all you have left are the student loans. Maybe it is time to change the game—when you next think that nobody really cares whether you're alive or dead, try missing a couple of student loan payments. You will quickly get a lot of attention.

Social Media, Emotions, Social Norms

ON KOPI LUWAK COFFEE

"Yes, it was tested on rabbits and they simply *adored* it."

Dear Dan,

During a recent trip to Los Angeles, I stopped by a coffee shop offering a very expensive coffee called kopi luwak, or civet coffee. I asked about the steep price, and the barista told me the story of the special process required to make this coffee: A catlike Indonesian animal known as a civet eats coffee cherries and then poops out what are basically beans. People then collect these "processed" beans and use them to make a highly unusual brew that's said to be smoother than its journey. It can sell for hundreds of dollars per pound. I was curious but not interested (or brave) enough to buy it—let alone drink it. Can you explain why are people willing to pay for this?

—CHAHRIAR

First, I think you made a mistake. You should have paid up and tried a cup—in part because you are still clearly curious about this unique and unusual coffee, and in part because it would have made a much more personal and interesting story (and what are a few dollars compared to a good story?). So next time you pass by a coffee place with kopi luwak, try it—maybe even get the double shot with hair and all the add-ons.

As for civet coffee's quality: The promotional material that I found says that civets know how to pick the best coffee beans and that their digestive systems ferment the beans, reducing their acidity and providing a much better coffee. (I have no idea how this works exactly, but the story is interesting.)

And the big question is, why are people willing to pay so much for civet coffee? One reason is that they are paying for the novelty and the story. Another reason is based on the amount (and type) of labor involved. This particular process of production is clearly much more complex than your average cup of java and, in general, we find that people are willing to pay more for something that required more effort to produce—even if the product itself is not better—and civet coffee sounds like a prime example of this effort-based-pricing principle.

Finally, I wonder how much people would be willing to pay had the beans passed through not an Indonesian animal but an American human. My guess is that despite being a very good story and despite the amount of effort that would likely be involved, this particular version of a brew would be too strong for us.

Food and Drinks, Value, Experiences

ON WEDDING
RING WOES

"If you're going to go down that road,
be prepared for two-way traffic."

Dear Dan,

My wife-to-be really wants to get a two-carat ring, but I'd rather get a smaller ring and use the rest of the money for future expenses—house, wedding, etc. Her view is that most of her friends have big rings, plus she's been dreaming about this for a long time. What do you think about this irrational behavior? Any advice?

—JAY

First, there is a difference between irrational and difficult to understand.

One way to view this desire for diamond rings is that women like these things exactly because men hate shopping for them. If you purchased something for your loved one that you actually enjoyed shopping for, this would be nice—but having to overcome your aversion to shopping for something you don't want is a much stronger signal of your love and care.

For example, let's say you purchased something for your loved one that you enjoyed shopping for or that you personally wanted—a new SLR digital camera, for example. While this would certainly be a wonderful gift and I am sure it would be highly appreciated, the problem is that it would be hard to tell how much of your effort is due to your romantic love and how much is due to your selfish desire for the gift itself. On the other hand, if you purchased something that you didn't like shopping for and even hated both the process and the product, your action would make it crystal clear that you are doing this solely because of your deep, romantic love and commitment to your significant other. This is why it is important to buy something you dislike and don't understand its value as a true signal of your love and care.

So, this year, when you are shopping for jewelry or flowers for your soul mate, remind her what a pain it was for you. And if you want to prepare for next year, you should start broadcasting how much you hate SLR cameras and how agonizing and time consuming the shopping process is for one of these useless products.

Relationships, Giving, Signaling

ON SOCIAL VIOLATIONS AND TATTLE-TELLING

Dear Dan,

I was the whistleblower for a very large corporate disaster. Since my whistleblowing, I have been shocked at the vitriol and social exclusion I have suffered as a result of speaking the truth. What is it about whistleblowers that makes society want to exclude us? Any insights and guidance would be most welcome.

—WENDY

From what I understand, the backlash you are experiencing is very common among whistleblowers.

As I was reflecting on your question, I started wondering why I want my kids (ages twelve and nine) to solve their problems themselves, without involving the higher authorities (their parents). For some reason, even with my kids, I view tattle-telling in a very negative light. Of course, sometimes my kids have legitimate grievances that require an intervention from the authorities, but my negative reaction to tattle-telling suggests that I'm willing to accept some violations of family justice in order to have their issues resolved internally.

Perhaps the friends and colleagues of whistleblowers see them as outliers of the social fabric—since they've shown willingness to seek external authorities when conflicts emerge. Maybe this social exclusion is due to a belief that when problems will emerge in the future, whistle-blowers will again look for an external authority. If you were Tom Sawyer, you could cut your hand and mix your blood with that of your friends to symbolize your connection and loyalty to the group, but given that this might not work for your age and social group, perhaps you need to find a related ritual that will demonstrate and solidify your deep commitment to your social group.

Workplace, Family, Rules

ON VARIETY AS A
MEMORY ENHANCEMENT

Dear Dan,

My best buddies and I have a tradition of going on a one-week ski trip once a year. We've been doing it for most of the past decade. The idea is that it's just us guys on the mountain, enjoying good company and snow. We cherish these moments and can't wait for the week to arrive every year. The problem is that once we land at our ski destination, time seems to go by at light speed. The week ends amazingly quickly and when we look back at our time together it seems even shorter. I know that "time flies when you are having fun," but is there a way to perceive the week as longer?

—AVI

Given the way you phrased the question, the answer is simple: Take your wives with you. (Sorry, I couldn't resist.)

But more to the point: I suspect that one of the reasons your vacations seem so short, both as you experience the week and in your memory after the fact, is that the days of skiing are so similar to each other that they tend to blend together into one long experience rather than seven separate days of vacation.

On your next trip, try to make the days more different from one another. Try snowboarding one day, take a day off from skiing in the middle of your trip, take a ski lesson on another day, go sledding, or maybe even just change your ski equipment. The point is that even if some of these activities end up being less enjoyable in the moment relative to your regular skiing, they will help you categorize your vacation as a series of distinct experiences instead of one prolonged episode of skiing. This way, you will get a larger variety of experiences and more appreciation for all that you accomplished during your amazing week with the guys.

Travel, Time, Experiences

ON THE BENEFITS
OF A CROWDED SPACE

"Fine. Sit there and check your messages. Perhaps it will give you something to contribute to the conversation."

Dear Dan,

Why do young people on dates go to loud, crowded places? The noise in these places must prevent the potential couple from talking to each other and it virtually eliminates any possibility that they will get to know each other. So what's the point?

—AMANDA

Have you considered the possibility that daters who go to bars, clubs, and other noisy places are not really interested in getting to know each other? Maybe they have a different goal?

More seriously, noisy and crowded places might seem an odd choice for a date, but in fact these environments might help daters in multiple ways. First, noisy environments can help socially clumsy interactions by masking awkward silences. Second, if the could-be-couple runs out of topics from time to time, they can have the illusion that the silence isn't due to their inability to keep up a lively conversation and can, instead, chalk it up to the difficulty of talking over the music or to their fascination with the music being played.

A third benefit of such venues is that the noisy surroundings can give the couple an excuse to get physically closer to each other in order to be heard. A particularly loud bar may even give them permission to talk directly into their date's ear. (Permission to softly blow into the other person's ear is optional.)

Finally, music and crowds have been found to be very effective in increasing the general arousal level. Yes, arousal. With noise and people all around them, daters are likely to feel a much higher level of arousal, and, most important, they may misattribute this emotional state to the person they're with. (Social scientists call this "misattribution of emotions.") To the extent that people confuse the emotions created by the environment with the emotions created by the person sitting next to them, going out to loud, busy places could be a winning strategy. Just imagine leaving a bar after two hours where, during the entire time, the person with you was certain that the strong emotional feeling that he/she experienced was all stemming from you.

I hope this explains the mystery—and inspires you to start going on dates in noisy places.

Relationships, Sex, Emotions

ON HIRING A GOOD (AND FREE) ADVISOR

Dear Dan,

What is the best way to inject some rationality into our decision-making?

—JOE

I am not sure about the best way, but here is one approach that I use from time to time and maybe it will be useful for you as well. When we face decisions, we often see the world from an egocentric viewpoint. We are trapped within our own perspective, our own special motivations, and our momentary emotions.

One way to overcome this perspective and look at the situation in a cooler, more rational, and more objective way is to switch our perspective and consider what advice we would give to our best friend if they were in the same exact situation.

For example, in one experiment we asked people to imagine that they have been treated by the same doctor for the past ten years, and that this doctor just recommended a very expensive treatment for them. We then asked them if they would seek a second opinion. Almost everyone said "no." We asked another group to imagine that a friend was in the same exact situation, and we asked if they would recommend that the friend seek out a second opinion. Most advisors said "yes."

These results suggest that when we have a longtime commitment to a particular doctor, it's hard to ignore this relationship and our feeling of obligation. But, when we think about giving advice to other people, we can disengage from our emotions to some degree, look at the big picture, and recommend a better course of action—such as getting a second opinion.

Taking this "advice approach" may not be the best way to inject rationality into our decision-making (and it's certainly not the only way), but I find it useful to imagine what advice we would give to another person, particularly someone we care a lot about.

P. S. This discussion on advice makes me wonder what advice I personally should take from these pages . . .

Decisions, Emotions, External Perspective

ON THE GARLIC EFFECT

Dear Dan,

My daughter recently persuaded me to start eating two cloves of garlic every day. As a result I now feel more energetic and less stressed. Is it the garlic, or is it a placebo?

—YORAM

I am not sure about the real power of garlic but have you considered the possibility that the reason you feel so much better is that the people around you are now leaving you alone?

Food and Drinks, Health, Other People

ON GIVING TO THE POOR

"Spare a little eye contact?"

Dear Dan,

I was recently approached by a panhandler who asked for 75 cents. I was late for my train, so I gave him the change I had in my pocket and hurried along. Only later I started wondering why he chose 75 cents. Do you think the 75-cent request could be a "market tested" amount, one that yields a higher overall level of donations than asking outright for a buck or more?

—BRAD

I am not sure if the panhandler came up with this strategy after substantial research or just based on intuition (if it was based on research, then he deserves more credit than most Fortune 500 companies), but for sure this strategy is interesting.

One possible reason this approach could work is that by making this unique request, the panhandler could be trying to separate himself from the competition, causing passersby to stop for a few seconds, look at the beggar, think differently about him, and maybe give him some money. Another possibility is that by stating an exact amount, the panhandler was able to change the inferences you were making about him and his situation. The idea here is that the granularity of the way we express ourselves communicates important information. For example, when someone tells us to meet them at 8:03, we come to a different conclusion about how serious they are about meeting at an exact time compared to someone who tells us to meet them at 8 or 8-ish. In the same way, a request for exactly 75 cents may carry with it a set of inferences about how seriously the person needs the money. This very specific request may lead us to think that there is a particular reason for the request—like getting money for bus fare—and we are more likely to help when the need is specific.

You could argue that the same principle would apply if he asked for $1.25, but in this case the size of the request might deter some people. Plus, when he is asking for 75 cents, and people don't have exact change, there is a good chance that they will give him $1 and not wait for change. But if the requested amount is $1.25 giving $2 might be too much.

Having said all of this, I should point out that my speculations about the specificity of the request and about the rounding-up effect are just speculations and the right thing to do is to carry out some experiments. So, if you are willing to volunteer and beg for money for a few days, we can see how people react when you ask for different amounts, give people different reasons for the amount requested, and make it easy for them to get change back. By gathering some real data we should be able to truly get to the bottom of this strategy.

Beyond gaining a better understanding of begging strategies, carrying out such experiments might be useful in figuring out how to ask people for help in our everyday life. It might help us figure out how to make people stop and listen, how to influence the inferences that they make about us and our requests, and ultimately it can help us get the exact assistance we need. Now that you see how important this is, let me know when you are ready to start the experiments.

Giving, Attention, Value

ON GRANDPARENTS
AND AGENDAS

"Can I call you back? I'm creating happy memories
of my childhood for my father."

Dear Dan,

For the past five years our daughter has been married to a wealthy, bullying control freak. We have no sympathy for her; she is an admitted gold-digger, and her husband has boatloads of money. We have not been able to hide our disappointment, and knowing how we feel about their marriage, they have shut us out. We don't care too much, but we would love to spend some time with our grandson. We were even thinking about a legal approach, but grandparents have no visiting privileges in our state. Any advice?

—REG

I am sorry about your predicament, and while it is hard to give advice on this complex issue, here are a few suggestions. First, try calling your daughter and her husband and simply say that you're deeply sorry about all your previous behavior and negativity. You don't sound sorry to me, but that's OK. Just say you're sorry and say it repeatedly. In multiple experiments, we found that saying sorry works rather well—even when people don't mean it. Most interestingly, an apology works well even if the person from whom you are asking for forgiveness knows that you don't really mean it.

The point is that when someone says he or she was wrong and asks forgiveness, it's hard to continue being mad at them. You might find it difficult to swallow your pride but think about the relationship with your daughter, her husband, and your grandson as a game of chess. You really care about the king (seeing your grandson), and pride is just a pawn in the game (well, maybe a bishop) so it's OK to sacrifice it for something you really care about.

If this approach doesn't work, either because you can't bring yourself to say sorry or because the hatred is too deeply rooted—

and if you're serious about spending time with your grandson—I would recommend that you take all your belongings and move in next door. This will force some interaction between all of you and, with direct interaction, hatred will be harder to maintain—particularly if you are nice to your grandson (what parent can hate people who are nice to, and care about their kids?). On top of that, if your grandson makes it clear to his parents that he wants to spend more time with you, what are the chances that the parents could say no?

Finally, I should mention that in my personal experience, not only is living next to my parents-in-law incredibly helpful, meaningful, and useful, but the pleasures of living next to my extended family has exceeded my expectations.

Family, Memory, Forgiveness

ON OUTSMARTING
BATHROOM GOERS

Dear Dan,

　　Whenever I go to a public bathroom, I try to figure out which stall I should use. Any advice?

—CATHY

I assume that your question is about which bathroom stall is likely to have been used the least. But what you are really asking about is the level of sophistication of your fellow bathroom goers, and how to outsmart them.

If we assume that other people don't think about this question and just pick a stall randomly, this is bad news for you because it means that all the stalls are going to be equally used and there is no way for you to beat the system.

On the other hand, if the common bathroom goer picks their stall deliberately, you can try to get into their head and strategize one step ahead of them. So, we need to figure out what the common bathroom goer thinks. If they think that the closest stall is the least used (maybe because people feel that the farthest stall is more private), they will flock to it, and the closest stall becomes the most used one. In this situation, your best strategy is to do the opposite of what most people do, and pick the stall farthest from the door. But what if the common bathroom goer is a bit more sophisticated? What if they come to the same conclusion (that most other people think that the closest stall is the least used and, therefore, flock to it)? Thinking this way, they will try to counteract this choice by picking the farthest stall. What is your best strategy in this case? Again, you need to stay one step ahead of the game. In this case it means that your best strategy is to grab the closest stall. Of course, figuring out what most people do and taking the opposite approach depends on how many steps ahead people usually think.

So, now we get to the most important question: How many steps ahead do people usually think? My own personal (and sad) observation is that people think ahead by about one step (sometimes less). This means that you should start with the assumption that most people think that the farthest stall is the most used one, do the opposite, and pick the closest stall. Which means that you need to do the opposite of the opposite and select the farthest stall.

If nothing else, I hope that this analysis will give you some appreciation for the complexity of making the right choice in a competitive environment and maybe it will give you something more interesting to think about on your next visit to a public bathroom.

Decisions, Other People, Predictions

ON GOSSIP AS A SOCIAL COORDINATION MECHANISM

Dear Dan,

I'm at a loss for understanding the popularity of gossip newspapers and magazines. What is the attraction?

—DAVE

I don't fully understand it myself, but I suspect that some of the attraction has to do with social coordination. When we are in social gatherings we look for discussion topics that everyone can take part in, and often these end up being about weather/sports/gossip. This, of course, also means that discussions naturally gravitate to some of the lowest common denominators—just so that everyone can be part of the exchange.

Even though sports and gossip don't require a lot of knowledge in order to join a discussion, they do require some. Knowing that at some point we will find ourselves in social situations that will turn to these topics for discussion, and wanting to fit in, we consume gossip and sports news just so that we can participate more fully in passing time together.

P.S. In *The Hitchhiker's Guide to the Galaxy* Douglas Adams had the following to say about the topic: "Nothing travels faster than light, with the possible exception of bad news, which follows its own rules."

Communication, Other People, Social Norms

ON FRIENDS
WITH BENEFITS

"If we're going to be friends with benefits,
I want health and dental."

Dear Dan,

My boyfriend and I have been together for a while, and people ask us whether we're going to get married. We get along great and love each other very much but I just don't see the point of marriage. Why not just live together in a civil union and be happy the way things are? Aside from the cost, is there any point to this elaborate ritual?

—JANET

I have no research on this topic, but allow me to share a story that might help you think about your question.

When I was about nineteen, I was moved from the burn department (where I was initially hospitalized in basic isolation) to a more general rehabilitation center. In this rehabilitation center I met patients with injuries that ranged from amputation to paralysis to head injuries. Among them was David, who was stationed in the army as an expert in explosives, and had been badly injured while disassembling a land mine. He lost one of his hands and an eye and also had injuries to his legs and some scars. When Rachel, his girlfriend of several months, broke up with him, all the patients in the rehabilitation center, myself included, were furious with her. How could she be so disloyal and shallow? Did their love mean nothing to her? Interestingly, David was better able to see her side, he was not as negative as the rest of us about her decision, and he was the only one defending her.

Looking back at it now, I am not sure if Rachel was right or wrong, but her behavior can help us reflect on your question. Think about her behavior. Does it upset you? How might your feelings toward her change if her relationship with David had been a longer-term relationship? What if they were engaged? What if they were part of a civil union? What if they were married? How would you behave if you were in Rachel's position in each of these different types of relationships? And how would you expect your significant other to behave if he were in Rachel's position in each of these different types of relationships?

I suspect that your level of scorn for Rachel depends to a large degree on the type of relationship between her and David. I also suspect that your predictions about your own decision to stick with a partner who just experienced an awful injury (as well as your prediction about what your partner would do if you were the one injured) would similarly depend to a large degree on the type of relationship. And what is the lesson from this? If your assessment changes when you imagine that David and Rachel were married, this suggests that publicly saying "for better and for worse" really means something to you. It also means that getting married could change the way you view your own relationship.

Obviously, marriage is not some magical superglue for relationships. But marriage can be an important catalyst for commitment and long-term relationships, particularly when we inevitably hit rough patches. So while I wouldn't advocate marriage in all situations, I do think it's worth thinking about the ways in which this tradition can strengthen the long-term bond between people.

Relationships, Long-term Thinking, Happiness

ON RUMINATING WHILE RUNNING

Dear Dan,
   I often hear people say that after they go for a run, their minds are clear, and they can focus better on big questions at work. Can this be so? Do we need to exercise to think clearly?

—SAM

I suspect that running isn't the best way to clear the mind. In fact, I suspect that running while thinking about work is a recipe for designing products and experiences that enhance agony, misery, and pain. Now that I think about it, maybe this was the start of what we know as customer service for cable companies.

Workplace, Experiences, Misery

ON THE JOY OF
GETTING THINGS DONE

"I suppose a fist bump would be more sanitary."

Dear Dan,

Why do I clean my cell phone many times a day and with great care but I can't be bothered to deal with the cleanliness of my car or my house?

—SARA

I suspect that this is about your ability to reach your end goal. You probably don't really believe that under any circumstance you will ever reach a state where your house is 100 percent clean. The task is just too large, and others in your household can mess the place up faster than you can clean it. Given this state of affairs, maybe 70 percent clean or 80 percent is within reach, but 100 percent is just not going to happen.

On the other hand, when it comes to your phone, perfect cleanliness is within reach and this achievable goal spurs you not only to clean the phone but also to enjoy it.

I encountered a related case awhile ago when we hired a contractor to renovate parts of our home. The renovations included many projects: replacing old windows, insulating the attic, installing a better heating system, renovating a bathroom, and installing a sauna in the basement. The whole process involved all the typical delights that accompany such projects: broken promises, annoyances, delays, unexpected surprises that ended up costing us more, changes to the original plans (that also ended up costing us more)—all the usual occurrences that a social scientist would naturally learn to expect.

The one thing I did not expect was the construction of the sauna. One evening the contractor asked me to join him in the basement. There he showed me with great pride how finely and precisely they cut the wood for the walls and the benches, how much care they took to make sure that the screws were bolted below the surface of the wood, and other such details. This was clearly not the usual level of pride that he showed in his work and it made me wonder about the joy of creating a complete thing. You see, all the other aspects of the renovation were only improvements, while the sauna was a complete stand-alone. As such, the sauna was potentially a perfectly constructed self-standing unit. Could it be that the potential to make something perfect increases our motivation? And could it be that when we are limited to just fixing something, our motivation is weakened? I suspect that this is the case, which means that maybe we should all start picking projects that are smaller, and more self-contained.

Habits, Effort, Goals

ON THE ART
OF MULTITASKING

"Are you multitasking me?"

Dear Dan,

I spend a lot of time on not-very-interesting conference calls using Skype and Google Hangouts. I usually try to answer emails during this time, so I turn off the video capability, so that no one can see me. On top of that I try to type quietly, so that no one can hear. But the sound of the keyboard seems to vibrate through the computer, and I suspect that the other participants know that I am not really paying attention. Any advice?

—KRISTEN

I suspect that you are not seeing the whole picture. The other people are most likely not noticing your typing because the sound from their keyboard overpowers yours. But if you are still worried about someone hearing your typing, get a tablet.

Attention, Technology, Workplace

ON CALLING
HOME

"When you head off to camp, your parents
will want to see some separation anxiety."

Dear Dan,

My son travels a lot and as a consequence we don't talk as much as we used to. Can you suggest a way that I can talk to him more frequently?

—YORAM

I suspect that your son has a very busy life and that his lack of calling does not reflect his love or level of caring for you. This said, maybe you can pick a regular day and time to talk, maybe even put it on your calendars, and this will make your conversations more frequent. And I promise to call both you and Mom the moment I get back from South America.

Love, Dan

Family, Time, Travel

ON TOASTS AND THE IDEAL SUPERSTITION

Dear Dan,

At a dinner party a few years ago, we were raising our glasses to our hosts' health. But, before we had a chance to touch each other's glass, the person on my right informed us that it is customary to look straight into the eyes of the person you're toasting as your glasses touch and that failure to follow this procedure will inevitably result in five years of bad sex. I don't think anyone around the table believed in the link between eye gaze and bad sex but we found it very amusing and, for the rest of the night, looked deeply into each other's eyes while toasting. I don't think of myself as superstitious, but since that dinner party, I find myself looking very intently into people's eyes when I toast. I know I am being irrational, so why can't I shake this superstition?

—KATHLEEN

If we were going to design an effective superstition, this one would be as close to perfect as we're likely to get. For starters, the cost of the ritual (looking into each other's eyes) is low, and in fact pleasurable. On the other hand, the cost of ignoring the ritual is very high (five years of rotten sex). It's certainly not worth risking such a large consequence for such a small, fun act. And like all good superstitions, the outcome in question takes place far into the future and is difficult to evaluate objectively. All of these are the foundations of a good, viral superstition.

The only thing I might add to the mix is a way to quickly fix the bad-sex karma in cases where someone mistakenly does not follow the ritual. Perhaps if someone forgets to make eye contact, they will have to close their eyes and have the person next to them help them drink the next sip of wine? With such an addition, we would have a truly perfect ritual and superstition.

Incidentally, I told a friend about this five-year deal, and his response was, "Only five years?"

Food and Drinks, Sex, Social Norms

ON PICKUP LINES
AND COMPLIMENTS

"If you could time-travel back to any period in history,
would your love for me keep you from going?"

Dear Dan,

I am happily married and was never much for the bar scene myself. But I do wonder if those cheesy pickup lines actually work. "If I told you that you had a beautiful body, would you hold it against me?" and so on. I can't imagine that anyone would buy such transparently empty flattery but these lines are so common that they must be doing something. Any insight?

—BARBARA

I'm no expert here, but my guess is that these kinds of pickup lines work much better than one might expect. There is some very interesting research showing that we love getting compliments (not a big surprise), that we are better disposed toward people who give us compliments (somewhat more interesting), and that we like those people even when we know that the compliments are insincere (which is the most surprising and interesting finding).

Beyond the insights into pickup lines, the implications of these results should get us to rethink compliments more generally. After all, compliments are free, they make the person giving them happier, they make the person getting them feel special, and they strengthen the bond between the two. So, why not just give more of them? With this in mind, try some pickup lines and compliments on your husband for the next few weeks, and let me know how it works for you, for him, and for your relationship.

Relationships, Appreciation, Predictions

ON THE ILLUSION
OF LABOR

Dear Dan,

Early in my career, I wrote a massive Excel macro for the large bank where I worked. The macro (a set of automated commands) would take a data dump and turn it into a beautiful report. It took about two minutes to complete the analysis and the report while an hourglass showed that the macro was working away. The report was very useful but everyone complained that the macro was too slow.

One way to speed up a macro is to make it run in the background, invisibly, with just the hourglass showing on-screen. I used this approach from the start, but just for fun, I flipped the setting so that people using the macro could see it do its thing. It was like watching a video on fast-forward: The macro sliced the data, different parts of the database changed colors, headers titles and graphs were created, and so on. The only problem? Now the macro took about three times as long to finish.

The surprising thing was that when I made this change, my coworkers were dazzled by how fast and wonderful the macro was. Do you have a rational explanation for this?

—MIKE

I'm not sure I have a rational explanation, but I have a logical one. What you describe so nicely is a combination of two forces. The first force is that when we are just waiting aimlessly we feel that time is being wasted, and we feel worse about its passage. This means that the time your coworkers were waiting aimlessly for the macro to do its thing was much more painful than time that was filled with some activity. The second force is that when we feel that someone is working for us, particularly if they are working hard, we feel much better about the whole thing. The point here is that we have a hard time directly evaluating the quality of the output that we get, but evaluating effort is something we do very readily and naturally. Interestingly, this joy at having someone work hard for us holds true not just of people but also for computer algorithms.

I hope that this experience taught you to use this lesson more generally in all your projects, and that with this insight about human nature, you will continue exposing your coworkers to more and more of the effort that technology does for them.

And beyond the lessons for the workplace, it should be clear that the same lesson also applies to your personal life. Whenever you can, work extra hard to describe to those around you how hard you have worked for them.

Workplace, Effort, Appreciation

ON MISERY AND
SHARED HUMANITY

Dear Dan,

I travel a lot for work, and I've been getting increasingly annoyed with the U.S. way of flying: the waste of time, the disrespect shown to passengers, and the lame excuses for delays that the airlines make right and left. Why are we putting ourselves in this horrible situation?

—DAVID

I'm not sure why we are taking this abuse lying down and not protesting more, but here's what helps me in the moments when I feel the same as you. First, every time I'm stuck on a runway, I try to think about the marvel of flight and remind myself how amazing this technology is. Second, I try to see the experience of travel misery as evidence of our common humanity. In my experience, security guards and airline staffers are just as rude and inconsiderate all over the world, suggesting that once we put people in similar situations (in this case, the same tiring, trying, and thankless service job), we all turn out to behave in more or less the same ways. In addition to serving as a vivid example of our basic and common human nature, my hope is that as more and more people travel and experience first hand our shared nature for themselves, we will come a bit closer to achieving world peace. Anyway, that's what I tell myself and, from time to time, it helps.

Misery, Travel, Waiting

ON FLASHY CARS

Dear Dan,

I don't care about cars, never have. But I'm a sales executive, and people tell me I should own a nice car (BMW, Mercedes,
etc.) to enhance my credibility with both my customers and sales
team. I can afford either but would rather save the cash and buy
a Honda. Does it matter?

—CODY

In essence your question is about signaling. The large and col-
orful tail of the peacock tells the peahen about his strength and
virility (if I can run around carrying this large and cumbersome
tail, just imagine how strong I am). In the same way, we humans
are concerned with the signals we send to those around us about
who we are. Signaling is part of the reason we buy large homes,
dress up in designer clothes, and buy certain cars. So the answer
to your question is yes. The kind of car that we drive communicates something about us to the world around us. Does it matter?
Yes again, because we are constantly reading these signals and
making inferences about their senders.

But some questions remain. What kind of signal do you want
to send? The BMW signal or the Prius signal? Maybe the signal
that you buy American-made? Maybe you want to get a clas-
sic car and show people that you take really good care of it (a
more subtle signal, but an interesting one). Another question
is whether the cost of the signal—in this case, the cost of the
car—is worth its signaling value. This depends on the nature of
the people you deal with, how well they know you, how often
you only have the chance to make a first impression, etc.

In the interest of full disclosure, I should say that I drive a
minivan. But now that I am thinking about it, maybe I should
stick a Porsche logo on it.

Cars, Spending, Signaling

ON DRESSING
DOWN

"It's gotten so customers won't take my advice
unless they think I'm gay."

Dear Dan,

I recently attended one of your lectures, and I was wondering why an Israeli guy telling Jewish jokes is wearing an Indian shirt?

—JANET

In general I am not someone who should be asked for fashion advice, but this particular case might be the one exception. I have a strong preference for being dressed comfortably, but the problem is that in many professional meetings there is a dress code that requires suits and uncomfortable shoes. I have no idea who invented this uncomfortable way of dressing, and I am almost sure that suits were invented and worn for the first time on a dare. But regardless of how suits were invented, here we are.

My solution? I figured that as long as I am wearing clothes from a different culture, no one who is politically correct (and this includes almost everyone in the United States) could complain that I'm underdressed. After all, any such critic could be offending a whole subcontinent.

Now that I think about it, maybe I should start giving fashion tips.

Fashion, Political Correctness, Happiness

ON EXPLORING
THE UNKNOWN

"Which trail has the best cell-phone reception?"

Dear Dan,

As summer finally gets closer, we are starting to plan our family vacation. For the past few years, we have spent two weeks vacationing in Florida. Should we stick to this familiar plan or try something different?

—MICHAEL

In general, sticking with something well known is psychologically appealing. Our attraction to the sure thing explains why, for example, we often frequent the same chain restaurants when we travel and even order the same familiar dishes and the same flavor of ice cream. Sure, we might enjoy something new more than the sure thing, but we might also hate it. And given the psychological principle of loss aversion (whereby we dislike losses more than we enjoy gains), the fear of a miserable experience looms heavy in our minds and we decide not to risk trying anything new.

That's a mistake for three key reasons. First, if you think about a long time horizon, say twenty more years of vacations and dining out, it is certainly worth exploring what else may be out there, what we love, and what experiences are best for us, before settling into a limited set of options. Second, variety really is one of the most important spices of life. Finally, vacations are not just about the two weeks you are away from work; they're also about the time you spend anticipating and imagining your trip, as well as the time after the trip when you get to replay special moments from your vacation in your mind. Among these three types of ways to consume the vacation—anticipation, the trip itself, and consuming the memories afterward—the shortest amount of time is spent on the vacation itself.

Given all this, the short answer is: try something new.

Travel, Experimenting, Happiness

ON TRYING OUT
RELATIONSHIPS

"I want to get married and start a family with you—
although God knows who I'll want to finish it with."

Dear Dan,

How can I decide if it is a good idea or not to marry my current girlfriend?

—NICK

In general, whenever you can, it is advisable to carry out experiments. This way you get quality data before making your decision. One of the key requirements for such experiments is to carry out the experiment in a setting that is as similar as possible to the situation you want to study. For example, if you want to study how people make decisions online, it is useful to get them to make the decisions on their computers, and if you want to study how people make decisions in a supermarket, it is good to place them in a hectic environment with lots of choices. What does this mean in your case? You are trying to understand how it would feel to be with this person in the decades to come—which leads me to recommend that you try spending two weeks with your girlfriend's mother.

Relationships, Experimenting, Happiness

ON DIVORCE AND
GOOD DECISIONS

"I wish you'd be more supportive of my efforts
to divorce you."

Dear Dan,

Why is the divorce rate so high?

—JACOB

It is hard to imagine we can be happy with any decision we make even one year down the road, much less when we look back at our decisions five, ten, twenty, or even fifty years later. Frankly, I am amazed by how low the divorce rate is.

....

Dear Dan,

I've been in a relationship with a girl for almost six years. The passion of those first days, when oxytocin levels were extremely high, is long gone. But I still feel comfortable with her. I don't know if it's time to leave or if I should stay and hope the passion returns?

—JD

It is hard to know what is the best thing for you to do given that I don't know anything about your age, your past experience with relationships, and what "comfortable" means to you.

This said, I suspect that you are experiencing the standard changes in relationships, where initial passion and attraction die off and are replaced by some other feelings ("comfortable" in your case). The question is whether "comfortable" is sufficiently positive for you.

When it comes to "comfortable," it is useful to consider the economist Tibor Scitovsky, who argued in his book *The Joy-less Economy* that there are two kinds of positive experiences—pleasures and comforts—and that we have a tendency to take the comfortable, safe, and predictable path way too often. This is par-ticularly sad, Scitovsky argues, because real progress—as well as real pleasure—comes from taking risks and trying very different approaches to life and types of experiences.

So, perhaps this is a good opportunity to give up your comfort and give pleasure a chance.

Relationships, Long-Term Thinking, Happiness

ON INVESTING IN
FINANCIAL ADVISORS

"You can't learn the path to financial independence
from just ANY infomercial."

Dear Dan,

Are financial advisors a wise investment? Mine charges me 1 percent each year for all my assets under management. Is it worth it?

—ALLAN

It is hard to know for sure. But the fact that many financial advisors have many different hidden fees suggests to me that they themselves don't think that people would pay as much for their services if they charged in a clear and up-front way.

It might help to think about this question in more concrete terms, and contrast two cases: In case one, you are charged 1 percent of your assets under management, and this amount is taken directly from your brokerage account once a month. In case two, you pay the same overall amount, but you send a check at the end of every month from your checking account to cover your financial advisor's monthly rate.

The second case more directly and somewhat painfully depicts the cost of your financial advisor, providing a better frame for the question of whether the cost is worth it. So, picture yourself in the second scenario, and ask yourself if you would pay your financial advisor directly for these services. If the answer is yes, keep your financial advisor; if the answer is no, you have your next action plan.

Money, Spending, Value

ON JUSTICE AND SHARING FOOD WITH SQUIRRELS

Dear Dan,

I find myself acting irrationally when it comes to squirrels. The rascals climb down a branch and onto my bird feeder, where they hang and eat like limber little pigs. When I see them in action, I rush outside yelling and I take great pleasure in frightening them away. But victory never lasts long. They come right back and the whole insane cycle starts over. My sister tells me that I need to watch *Snow White* again, to be reminded that squirrels are also part of nature, and that they are not inherently less worthy than the birds I so clearly prefer. Perhaps she is right but I am unable to embrace this perspective. Can you help explain what's going on with my reasoning, and how I might make peace with the furry marauders in my yard?

—NEARLY ELMER FUDD

It sounds to me that the root of your problem is that you view the squirrels' behavior as immoral. After all, in your mind the food was placed in the bird feeder for the benefit of the birds, and the squirrels are simply stealing it from its rightful owners. If this is the essence of the problem, the answer is very simple: All you need to do is to start calling the contraption a "squirrel and bird feeder." With this new framing, the squirrels will just be partaking in a joint dining experience, your problems should go away, and you might even be able to market this new product.

P.S. After I published this advice I received many passionate responses. These included suggestions for different contraptions, discussions of the overall immorality of squirrels, comparing the fight against the squirrels to the fight of Don Quixote, and even a detailed analysis of the financial losses to the U.S. economy due to these creatures. Clearly, there is much more to learn and explore when it comes to squirrels.

Food and Drinks, Morality, Giving

ON SOCIAL LIFE
AND THE INTERNET

"Don't look. It's the people we steal Wi-Fi from."

Dear Dan,

What is it about Internet communication—Facebook, Twitter, email—that seems to make people descend to the lowest common denominator?

—JAMES

It's easy to blame the Internet, but I think we see such behavior mostly because we generally gravitate toward trafficking in trivialities. Consider your own daily interactions as an example. How much is witty repartee—and how much is the verbal equivalent of cat pictures? The Internet just makes it easier to see how boring our ordinary interactions are.

Technology, Social Media, Coordination

ON EXPECTATIONS
IN DATING AND HIRING

"Just for future reference, save your private heartache
for the third or fourth date."

Dear Dan,

When it comes to job applicants, do we like people more or less when they're hired from the outside compared to when they are promoted from within?

—JOHN

Jeana Frost, Mike Norton, and I carried out a set of studies sometime ago showing that when it comes to dating, knowing more about a person leads to less love—not more. The basic finding is that when we know very little about potential romantic partners, our imagination fills in the gaps in overoptimistic ways (if he likes music, he must like the kind of music I like and not seventeenth-century Baroque) and then we meet them for coffee and our high hopes are crushed. Oddly, we also found that this disappointment shows up time after time, and that online dating enthusiasts don't seem to learn from their negative experiences, and don't tame their overoptimistic expectations.

Plenty of lessons from the romantic realm apply to other areas of our lives, and job applicants is one of them. There is some evidence suggesting that CEOs hired from the outside get paid more than those hired from the inside and that they don't perform as well. I suspect that the reason for this is the same heightened expectations that come with lack of knowledge—when someone is relatively unknown we tend to fill in the gaps in overoptimistic ways, we get more excited about their potential and as a consequence we are more likely to hire them and to pay them extra. But with hiring CEOs, the consequences of acting on the outline of our expectations is much more costly than a wasted hour and a cup of coffee.

Workplace, Relationships, Predictions

ON LEARNING TO BE BETTER DECISION MAKERS (MAYBE)

Dear Dan,

    Given all your research on decision making, and the mistakes we all make, do you now find yourself making better decisions?

—ODED

Maybe. Possibly. Sometimes. I suspect that studying the dark side of how we all make decisions helps me reflect on human decision making in general, but I doubt that it has any positive impact on the quality of my gut intuition. This means that when I rely on my intuition and gut feelings for making decisions, I am just as prone to mistakes as everyone else.

    Where I might do better is when I get to carefully consider my decisions. In these (rare) cases, the thought process is more deliberate, and under these conditions I might be able to avoid some of the decision traps that I know so much about. At least I would like to think that this is the case.

    Another benefit of understanding failures in decision making comes from recognizing the importance of habits. Habits are automated ways of acting without thinking very much, which means that to the extent that we create good habits, they can facilitate better behaviors. Given this, I try to delegate some of the most challenging decisions (overeating, under-saving, texting while driving) to rules and habits and I think this has worked very well for me so far.

I should also point out that much of my research starts with my observations of my own irrationalities, so, without my own mistakes, I would have to look for a different career.

Decisions, Long-Term Thinking, Habits

ON THE POWER
OF EXPECTATIONS

"I can't decide if that was bad in a good way,
good in a good way, good in a bad way, or bad in a bad way."

Dear Dan,

A lot of research in social science has shown that when we expect an experience to be of a certain quality, the power of expectations can alter the experience and, indeed, make it comply with our expectations. For example, your own research has shown that a glass of wine will taste better after reading a positive review of it, and that when we think that a beer will taste disgusting because it has some balsamic vinegar in it, our expectations will make us hate that beer (when in a blind taste, balsamic vinegar actually makes it taste better). Well, these findings mostly fit with my own experience; however, what you didn't mention is the possibility of a negative effect for expectations that are too good. In other words, is the effect the same when something is extremely overhyped?

My own observation is that when I passionately recommend a movie to my friends, sometimes their feedback is, "It wasn't that good. I thought it would be really amazing." I suspect that in those cases my friends are not experiencing the movie in the positive way that they should because I overhyped it. Do you think that overhyped expectations can backfire?

—OMID

My intuition is the same as yours. When I hype, or overhype something, my friends end up with very high expectations. These very high expectations become the standard from which my friends evaluate the experience—and when it inevitably falls short, their overall enjoyment of the experience is decreased.

Here is how I view the issue: Heightened expectations can change our experience by, let's say, 20 percent. This means that as long as the increased expectations are within this moderate range, the expectations can "pull" the experience toward them and influence it accordingly. But, when the expectations are too extreme (let's say by 60 percent or more), the gap between expectations and reality becomes too wide to bridge, and now, the contrast causes the expectations to backfire and reduce the enjoyment.

What I suggest is that if you want your friends to experience something as being better than it truly is, go ahead and exaggerate. But, not by too much.

Expectations, Happiness, Entertainment

ON COMMUNICATING
SAFETY

Dear Dan,

I have sometimes found myself walking behind a woman at night in a somewhat unsafe place, going in the same direction. Even though there is some distance between us, I can feel the doubt and worry in her mind. How should I handle this situation? Should I say something? I also need to be somewhere, but I don't want the woman to feel unsafe, so should I stop and wait?

—STEVE

No need to stop. Simply pick up your cell phone, call your mother, and talk to her in a slightly loud voice. In the world of suspicion, nobody who calls his mother at night could be considered a negative individual.

Other People, Emotions, Communication

ON THE
PERFECT GIFT

"Even if they don't like it, they'll be flattered
that we thought they'd like it."

Hi Dan,

Every year it's the same problem: My husband and I struggle to get his dad a few perfect gifts, only to see them sit unused for eternity. These are expensive and high-quality gifts—specialty tools for his car, toolboxes, super-handy gadgets, etc. But years later, the tools sit there unopened and the toolbox accumulates dust. He still carries his broken wrenches and stripped screwdrivers around in a ripped plastic sack!

Since the objects were "ours" at one point, we feel that we still retain some residual interest in what happens to them. Is it because we invested so much thought and effort in acquiring them? Is it because not using the gifts seems wasteful? Is it because we feel that the lack of use reflects badly on us? And now to the main question: Would it be so wrong if we just took our gifts back? He clearly doesn't want them and we could use them ourselves.

—VERONICA

No, you may not take the gifts back. (Note that I didn't write "your gifts," because I don't think you should consider them yours.)

The sad thing is that you and your husband feel unappreciated because your thoughtful and expensive gifts are not bringing the dear old man the happiness that you hoped to give him. Instead of taking the gifts back, I would try to increase the likelihood that the tools will get used. First, I would take them out of their packaging (which in some cases is so difficult that you need special tools just to open the package), and replace the old tools in his plastic sack with the new ones—thereby making the act of using the new tools easier, and more likely. As for the old tools, just put them in the attic for now, ideally behind some large boxes.

If your father-in-law protests, I would restore his old tool kit and suggest spring cleaning, including a donation of unused household goods to a local charity. He might be willing to give the new tools up for a good cause. And if that doesn't work, stage a robbery and steal the tools, leaving cash and other valuables untouched. The added benefit of the robbery approach is that it might also show your father-in-law how valuable your gifts are—and he might view your future gifts with new eyes.

As for this year, buy him something that gets better over time, such as good whiskey or wine. That way, if he doesn't use it, at least it will increase in value and bother you less.

Giving, Family, Relationships

ON EATING LESSONS
AND KIDS

Hi Dan,

Let's say you're very hungry and you plan to eat two sandwiches. One is very delicious and the second isn't as good. Which one should you eat first?

—PABLO

One of my college friends had kids many years before anyone else in our social group was even considering children, and he used to give the following advice (mostly unsolicited): "Think," he used to say, "about how you like to eat. There are some people who like to eat reasonably good food three times daily, while others would prefer to save their money and eat mediocre food most of the time but occasionally have an amazing meal.

"If you're the second type, go ahead and have kids, because life with kids isn't all that fun for the most part, but from time to time they bring incredible joy. But, if you identify with the first type, you may want to rethink having kids."

Now, I am not sure that anyone should use this metaphor when making decisions about having kids, but I do think that it works well for your question.

This thought experiment asks whether you are the type of person that focuses on the maximum amount of pleasure in any given experience, or if you are more concerned with avoiding the low points—the minimum levels—of your experiences.

One more element, and then we can get to your question: Let's also consider diminished sensitivity, which means that for any given dish, the first bite (when you are hungrier) is the best, that the second bite is slightly less good, and that the last bite will give you the lowest level of joy. (As Cervantes wrote in *Don Quixote*, "Hunger is the best sauce in the world.")

Now to your question: If you are the maximizer type, you should eat the better sandwich first so that the height of your initial joy comes from the combination of your hunger and the superior quality of the good sandwich. Of course, by taking this approach you will sacrifice the pleasure at the end of your experience but if you are the maximizer type, you should consider this a worthwhile trade-off. On the other hand, if you are of the avoiding-minimum type, and all you want is to get a more even experience—giving up the highs, but also eliminating low points—eat the so-so sandwich first. This way the initial part of the experience will be enhanced by your hunger rather than by the quality of the sandwich, and the latter part of the experience will benefit from the sandwich quality but take a hit from being late in the meal.

Personally, I prefer to focus on the most joyful part of the experience and eat the best sandwich first, ignoring folk wisdom to "save the best for last." Plus, this way I might be less hungry by the time I get to the so-so sandwich and eat a bit less.

Food and Drinks, Experiences, Happiness

ON USEFUL
COMPLAINING

"I thought we had the sort of relationship where
'please' and 'thank you' were implicit."

Dear Dan,

I recently met up with an old friend whom I hadn't seen for a very long time. I had been eagerly looking forward to our lunch, but I left very disappointed. All she did for more than two hours was complain—mostly about her husband, with some breaks to complain about her kids. It was a negative and depressing meeting and I left feeling bad for her, myself, and the time we spent together. Why do people complain so much? Could she really imagine that this was a good way to spend time with an old friend?

—ANDREA

People complain for a few reasons, and it is interesting and useful to figure out the exact reasons that our friends complain. One main reason is that misery often makes us feel closer to one another. Imagine that you are meeting a friend and you either tell her how difficult your husband and kids were last night, or give her the same level of detail about how wonderful your family is and how lucky you feel about the familial bliss you experienced yesterday. Under which case would your friend like you more, under which case would she share more with you in return, and under which case would you feel closer at the end of the meeting? I am willing to bet that it is the complaining one.

Another important reason for complaining is that we often look for reassurance—hoping the person we complain to will tell us that everything is OK and that what we're experiencing is just part of life. In fact, often what we really hope for is that the other person will share their own horror stories with us, that our experiences will pale in comparison and make us feel much better.

Now, let's return to your friend and ask why she was complaining. If she was looking to reconnect through shared misery, perhaps you should have indulged her efforts to strengthen your social bond. Perhaps you could have assured her that your bonds are strong and not in any need of additional strengthening. On the other hand, if your friend was looking for an emotional boost, maybe you should have told her something like: "You think your husband is a schmuck? Let me tell you about my prize." Thereby assuring her that her life is actually more normal than she might think.

Either way, complaining can actually be pretty useful. The next time a friend starts complaining, figure out the reason for it, and try to make the most of it.

Friends, Misery, Communication

ON PRICES AND
BIDDING FRENZY

Dear Dan,

My parents are about to put their house on the market in Scotland. In the Scottish system the seller sets an asking price and interested parties make a one-time, sealed-bid offer. Once all the offers are gathered the seller picks one and the transaction is carried out. Any advice on how to get the highest sale price?

—MOSES

Sealed-bid auctions are simpler in many ways than live auctions, and they involve two basic forces: what the bidders think the house is worth for them, and how intense they think the competition will be. Establishing a high asking price has an opposite effect on these two forces.

If you set a high asking price, there's a good chance that people will start thinking about the house in the price range of the asking price and offer a higher bid. On the other hand, if you set a low asking price, more people will participate in the auction, the competition will be fiercer, and the outcome is likely to be a higher final price. (By the way, have you noticed that in auctions—on eBay for example—the person who pays for the item at the end of the auction is called "the winner"? This suggests that competition is indeed a very strong driver in auctions.)

Now, the question is which of these two forces (starting perspective or competition) is stronger. I suspect that when the mechanism is a one-time, sealed-bid auction the most important element is the way people start thinking about the house, which suggests that you should go in with a high asking price. However, if you were in the United States, where the bidding mechanism involves multiple rounds, competition might be more important, which means that you would be better served by setting a lower asking price and getting more people to join the auction.

P.S. Last week I met with a friend who is house-hunting in San Francisco. He told me that the houses he has been bidding on had a very intense competition that ended up with selling prices that were about 30 percent to 40 percent more than the asking prices—a process that frustrates potential buyers. This brings me to my final point: A bidding frenzy might be good for a seller but since we are all going to be buyers and sellers at some point, it's not clear that the overall market for housing is better off with intense bidding frenzy.

Decisions, Value, Other People

ON TRANSMISSION
OF STRESS, AND CARING
FOR THE OLD

Dear Dan,

As a university professor who has been teaching for a long time, what advice would you give to students who are starting their academic year?

—PETER

Simple: Cut all ties with your family—particularly your grand-parents.

Here's why: Most professors discover that family members, particularly grandmothers, tend to pass away just before exams. Deciding to look into this question with the kind of rigor that only academics are able to (and have the time for), Mike Adams, a professor of biology at Eastern Connecticut State University, collected years of data and concluded that grandmothers are 10 times more likely to die before a midterm and 19 times more likely to die before a final exam. Grandmothers of students who aren't doing so well in class are at even higher risk, and the worst news is for students who are failing: Their grandmothers are 50 times more likely to die as the grandmothers of students who are passing the class.

The most straightforward explanation for these results? These students share their struggles with their grandmothers, and the poor old ladies prove unable to cope with the difficult news and die. Based on this sound reasoning, from a public policy perspective, students—particularly ones that are failing—clearly shouldn't mention the timing of their exams or their academic performance to any relatives. (A less likely interpretation of these results would be that the students are lying, but this is really hard to imagine.)

Kidding aside, social relationships are very important for our health and happiness, in good times and bad. And fostering these bonds is a wise goal for anyone at any stage of life.

Family, Procrastination, Morality

ON LUCK AS A MULTIPLE-STAGE NUMBER GAME

Dear Dan,

Are there people who are just lucky? I think so. Only I'm not one of them.

—AMY

Some people are indeed luckier, but it's not the kind of luck that gets you more money at the roulette wheel. Luckier people tend to try different things more frequently, and by trying more often they also succeed more. As an example, think about a basketball player who only shoots when he is 100 percent certain that he will make the shot. With this strategy he shoots three times in a game with perfect accuracy (3 baskets, and 100 percent success rate). Now, compare this to a player that tries 30 times but with a 50 percent success rate. With this strategy the second player will have 15 baskets and many more points.

On top of that, life is different from basketball in some very important ways. In basketball, every outcome is binary. A shot is either in or out. But in life, decisions often involve multiple stages and we can decide to try something and see where it goes. For example, we can start studying something new (go on a date, try a new food, interview for a new job, etc.), see if it fits our interests and skills—and only then decide whether to go deeper into the topic. This means that luckier people don't just try more things to start with; they are also quicker at cutting off the paths that don't seem to work out and focus on the more promising avenues.

So, what's the advice? First, life, to some degree, is a numbers game so try more things more frequently. Second, keep on examining all the options you have, and quickly cut the less promising ones in order to free more time for exploring options that might be better for you.

Luck, Experimenting, Decisions

ON SOCKS AND
THE PSYCHOLOGY OF
THE SUPERNATURAL

Dear Dan,

Why do socks always get lost in the laundry?

—JAMIE

Some time ago Ornit Raz and I looked into this fascinating question and discovered that otherwise reasonable people, who view themselves as having a strong grasp of the forces of nature, find themselves at a loss when it comes to this universal puzzle. The socks mystery often shakes people's faith in the laws of physics, and pushes even the skeptics to start believing in the supernatural.

We also found one psychological mechanism that can help us understand this mystery. The overcounting of missing socks. Most of us have many socks, and if we see one of them and don't immediately find its partner, we say, "Oh! A sock has been lost!" We remember that a sock is missing, but don't recall exactly its type or color. Later on, we see the matching sock, but we don't remember that it is the one who could form the pair with the first sock, so we say to ourselves, "Another sock is missing. Where is its partner? I can't believe so many socks go missing."

At the end of the day, the socks mystery is not due to the suspension of the law of physics. It stems from a much larger puzzle of how our memory works (or doesn't work).

At the same time, even with this scientific explanation, I still feel that at the back of my washing machine, there must be a black hole that is suitable just for socks.

Attention, Memory, Mistakes

ON TITHING

"I'm in the market for an easier religion."

Dear Dan,
Should Jews tithe?

—O.

Super-simple answer: Of course! Everyone should.

And obviously the best way to donate your money is by giving it to university professors so that they can continue with their important and illuminating research.

More seriously, giving money away is one of the most misunderstood human activities. We often think that if we have some money, the best way to use it is to spend it on ourselves. But there is a lot of research showing that giving money away leads to higher levels of happiness than spending it on ourselves. Of course, I am not recommending giving away all your money, but somewhere in the vicinity of tithing is a useful guideline for increased life satisfaction and happiness.

In addition to the general benefits of giving, specific rules such as tithing are very useful because they are strict and clear. When we have fuzzy and ill-defined rules (I will eat better, spend more time with my kids, drink less) it is easy not to think carefully about our behavior and about whether we are sticking to our declared goals or not. Unclear rules let us rationalize our misbehavior while keeping the hope that we will behave better in the future. In contrast, rules that are clear and strict (I will cut out desserts, read to the kids every night for thirty minutes, have only two glasses of wine per week) keep us from fooling ourselves and increase the chance that we will behave in accordance with our long-term best interest.

Tithing also helps in another important way. It changes our mindset from how much to give to where to give. When we tithe, the overall amount we give is a function of our income, and the amount is out of our hands (although I am never sure if tithing should be calculated before or after tax). Since the amount is set, we only have to decide where we want to have an impact. This makes giving feel a bit like giving away someone else's money—which is much more fun and rewarding.

Go forth and tithe.

Rules, Giving, Happiness

ON MIDLIFE
CLICHÉS

"I've had those books for years. They represent the
person I once aspired to be."

Dear Dan,

I am a middle-aged guy who's doing OK financially, and I'm thinking about buying myself a sports car. Perhaps a Porsche 911. But I'm also a bit disturbed by the obvious midlife cliché. What would you do?

—CRAIG

Tesla designs cars for people with your exact conflict. The Tesla is a sports car, but it has an environmental image and those who buy it can think of themselves as green, not gray.

Self-image, Cars, Aging

ON CHEATERS
AND ALIBIS

Dear Dan,

I recently stumbled upon a website offering customers help with creating alibis. It even manufactures corroborating "evidence" for their absences (for example, it reassures your wife that you were at a conference when you were really with your mistress). Other sites offer married people help finding paramours for extramarital affairs. Do you think these sites are increasing dishonesty?

—JOE

Thanks for your question. In addition to being interesting, it also led me to explore some of these websites myself. And the basic answer to your question is "Yes." I think that these websites do increase dishonesty.

As far as I can tell, many of these websites are constructed to look as similar as possible to websites of more mandarin types of services. In one case, I saw a website featuring pictures of smiling people wearing headsets, waiting to fill an order for services ranging from producing and sending fake airline tickets, to impersonating hotel reception. The testimonials on another website were very positive and very general, and yet another website included the slogan "Empowering Real People in a Real World!" which could be downright uplifting, until you realize that when they say "empowering" people, they mean lying on their behalf.

I suspect that this type of phrasing and suggested commonality help people to rationalize their actions as socially acceptable. And with all the testimonials from so many regular people, why not you?

I also think that the "real world" rhetoric may further lull people's objections by promoting the idea that this is how things truly work in the real world as opposed to the fairy-tale land of perfect honesty that some people pretend to be part of.

For my part, I'm left feeling a little worried about what kinds of ads might pop up in my browser after looking at all of these websites.

Relationships, Honesty, Technology

ON BREAKFAST
REGRETS

Dear Dan,

I often buy a breakfast sandwich from my regular café. Some-times, I take the empty paper wrapper, walk five meters to the trash bin, dispose of the wrapper, and walk back to my seat—a perfectly convenient sequence of events. But at other times, without getting up from my seat I try to throw the wrapper into the trash. I am a lousy shot, and when I (inevitably) miss, I have to make the same journey back to the bin, pick up the wrapper, and place it in the trash bin. The trip is the same in both cases, but walking to the trash bin after I have missed feels much more like a chore.

Why do I feel so differently about the same journey?

—RICHARD

The answer to your question lies in the realm of counterfactuals, which is thinking about what could have been and comparing what we have at hand to that alternate reality. Here is how coun-terfactuals translate into your daily drama: When you aim and miss, you can clearly imagine a world in which you sank your shot. You judge your efforts by comparison to that imagined world, and, in relative terms, you feel bad about it. But when you don't even try to make the shot, there is no other world to imagine and no contrast to make you feel bad.

My suggestion: Buy your sandwich and order your coffee, but ask the café to make you the coffee three minutes later. Go to your table, sit with your sandwich, and try to shoot the wrapper into the trash can. Now, no matter how successful you are, get up and walk to the counter to get your coffee. If you made the basket, great; if not, pick up the wrapper on your way to get your coffee. This way there is no world in which you didn't have to get up after your shot, no counterfactuals, and no comparison to make you feel bad. Happy breakfast.

Regret, Emotions, Food and Drinks

ON NIGHTTIME ACTIVITIES

"If only we could stay home and TiVo the Carlsons."

Dear Dan,

My husband and I are childless. We've lived in the same house in the same town for seventeen years. Each day my husband comes home and says, "What do you want to do tonight?" By now we've tried every restaurant in a five-mile radius so often that we almost know the menus by heart. Neither of us enjoys shopping or watching movies at a theater. His hobby is aviation, and I have an aversion to flying. I work from home and would love to go out in the evening occasionally, but we usually end up just staying at home and watching TV. And we don't even like TV! Can you shed some light on this problem and suggest how we can get out of this rut?

—CHARLEEN

The basic challenge you are facing is what economists call a problem of coordination. Every night you and your husband look for an activity that you can agree on and that both of you will enjoy. This is no easy task when your preferences for the ideal activity don't align. On top of that, you have the suboptimal default option of watching TV—something that neither of you enjoys but is a simple solution to your coordination problem, and something you can easily fall back on whenever you can't reach a better solution.

One approach to your conundrum is to switch it from a simultaneous coordination problem to a sequential one. For this approach you will need to agree up front on a plan that will make only one of you happy on any given night but, ultimately, will let both of you experience activities that you enjoy to a larger degree. Here are some practical steps you can take in order to set up this sequential coordination: Using some cards, write down activities that you want to engage in, and ask your husband to do the same using the same number of cards. Mix the cards and every evening when you don't know what to do, draw one card to pick that night's activity. You also have to commit in advance that when these cards are drawn you will follow up with the named activity. You can easily see how using this approach would make one of you very happy (the person whose activity was chosen) and it should also lead to higher overall enjoyment. After all, it's better to have a high level of enjoyment on some nights of the week than to have no joy every night.

And one final suggestion: Add a few wild cards to the mix (singing, poetry, pottery, volunteering, square dancing, etc.)— activities that you aren't sure if either of you would like. If you follow this strategy, some nights will end up being an unpleasant learning experience—giving you new insights into the extent to which you dislike these activities—but there might be some evenings where you will be surprised to find some new activities that you both truly enjoy.

Relationships, Coordination, Experiences

ON PLAYING
PARENTS

"We planned on having you but you're not the children
we planned on having."

Dear Dan,

My wife and I are in our late thirties, and we are debating whether or not to have kids. Any advice?

—HENRY

The decision on whether or not to have kids is very complex. It depends on many factors, including your financial situation, your preferences, and the quality and stability of your relationship. So, sadly, without knowing much more about your situation I can't provide a direct answer to your question.

At the same time, given that this is one of the most important decisions you will ever make, I feel some obligation to point out some general lessons that apply to large and substantial decisions.

Like many other decisions, here too the question is all about what you might get from this experience and what you might have to give up. The problem is that before you have kids, it is hard to estimate both the costs and the benefits. So what should you do? You need to try to simulate the kid-having experience in order to have a better understanding of what it means and how it would fit your preferences and life.

To get more insight into this question, why don't you, for example, move in for a week with some of your friends who have kids and observe them up close? Next, why don't you offer to take care of some other friends' kids for a week? Then try to expand this exercise and take care of kids from different age groups (don't skip very young kids and teenagers). After ten weeks of these types of experiments, you should be in a much better position to figure out if this particular activity is right for you or not.

If this exercise seems too daunting for you, you probably fall into one of two categories: 1) You're not really that interested in an empirical answer to this question. Perhaps you've already made up your mind and it is just that you're not yet ready to admit it. 2) You're too lazy to put the effort into figuring this out. And if that is the case, you probably should not have kids.

Experimenting, Family, Happiness

ON JOINT
ACCOUNTS

"You should try taking more naps yourself. Sleep is free."

Dear Dan,

I recently got married, and my wife and I have been debating the topic of bank accounts. She'd like to combine our bank accounts, because she wants to know how much is coming in and going out. I think separate accounts would be simpler for taxes, personal spending, and budgeting. What's your take?

—JONATHAN

The fact that you're wondering whether to follow your own or your wife's ideas on what is the best way to run your household tells me that you are either a slow learner or very recently married (sorry, my Jewish heritage would not let me pass up this opportunity). But to your question: I think you should have a joint account.

First, there's no question that in reality your accounts are joint in the sense that anything one of you does has a direct effect on your mutual financial future. For example, if one of you starts buying expensive cars from your individual account, there's going to be less money for both of you to spend later on vacations, medical bills, and so on.

Beyond the legal point and most important for the whole marriage enterprise is the fact that by getting married you have created a social contract in the form of "I will take care of you, and you will take care of me." The mutuality of this agreement is a crucial key to the success of any marriage, and adding a layer of financial negotiations to this intricate relationship can easily backfire.

For example, think about what would happen if there were "my money" and "your money." Would you start splitting the bill in restaurants? What if one of you has an extra glass of wine? And what if your wife runs out of "her money"? Would you tell her that if she does the dishes and takes the garbage out for a week, you will give her some of "your money"?

The problem is that once money becomes intertwined in deep social relationships, the relationship in question can start looking a bit more like prostitution than love, romance, and long-term caring. Separate bank accounts will certainly have some financial advantages, but separate accounts could also put unnecessary stress on your relationship and I hope it's clear to you that sacrificing some efficiency for the sake of a good relationship is a worthwhile tradeoff.

Relationships, Money, Social Norms

ON THE BORDEAUX BATTLEFIELD

"I'll give you a few moments to recover from the prices."

Dear Dan,

I love dining out, including having some wine with dinner—but the truth is that I can't tell much difference between different bottles, and I never know which wine to order or how much to spend. When I ask waiters or sommeliers for advice, they often give some flowery descriptions about soil and accents of apricot, but these never help me figure out which wine pairs best with my meal. The whole wine-ordering business makes me feel incompetent and inadequate. Do you have any simple advice for how to order wine?

—JOSH

The first thing to realize when picking a bottle from a wine list is that you are on a battlefield. And this is not a regular battle. It is a battle for your money. A battle between the restaurant (which wants as much of your money as possible) and your savings account. And what's worse, the restaurant's managers have much more data than you do about how people make their decisions (including decisions about wine), and they have the first-move advantage by setting up the menu in a way that gives them the upper hand.

In particular, restaurants know that people make relative decisions, which means that if the wine list includes some very expensive wines (say, bottles for $200 or more), customers are unlikely to order these very expensive bottles, but their mere presence on the list will make a $70 bottle seem much more reasonable.

Restaurants also know that many of us are cheap but we don't want to seem cheap, which means that almost no one orders the cheapest wine on the menu. Instead, the wine of choice for cheapskates is the second-cheapest wine on the list. Knowing this, the restaurant places a wine with a relatively high margin in this attractive location on the wine list.

Finally, the restaurants have another weapon in their arsenal: waiters and sommeliers who add to our feelings of inadequacy and confusion and, in the haze of our decision making, can easily push us toward more expensive wines.

Now that you are getting the picture, and are thinking about the ordering of wine as a battle, you can think ahead. Perhaps you can decide in advance to spend up to a certain amount of money on wine. Or possibly tell the waiter that you have a religious rule against spending more than a set sum on wine and ask for a recommendation that would fit within your religious boundaries. If you really want to strike back, inform the waiter that you have allocated a total of fifty dollars for the tip and wine combined—so the more you spend on wine, the less you will leave as a tip. Now let's see what they recommend.

Food and Drinks, Spending, Decisions, Value

ON TRAFFIC JAM
ALTRUISM

"That other driver got the thank-you wave that I deserved."

Dear Dan,

Often as I creep along in a traffic jam, someone inevitably tries to enter my lane from the side. Now here is the issue: If I let the car in, I feel good about it. But when I see others in front of me let someone in, I feel cheated, because I've been waiting longer than the car entering the lane, and I am upset with the driver who acted kindly at my expense. Any idea why I feel so different about these two situations?

—WALT

The issues here are control and credit. When you let someone into your lane, you're the one making the decision and you're the one getting the nod or the hand wave as an expression of gratitude. In contrast, when someone else is letting the needy car in, you have no control over the decision, and you're not getting the credit. You only see the downside of this action in the form of an increased delay.

This analysis suggests that your emotional reaction is not just about the other driver. To illustrate this point consider a case where it is not another driver who is doing a favor to a car trying to join your lane. In this imaginary setup you simply keep a large distance between you and the car in front of you. By doing this, you're allowing the cars from the merging lane to come into your lane at will, but it doesn't require a separate act of generosity on your part. You don't even need to slow down to let the car in. My guess is that this version of helping other drivers also wouldn't feel very satisfying for you, not to mention that you're not going to get any credit for your passive kindness.

What's the conclusion? First, it is not about the other kind driver. It is about you! Second, to feel good about the good fortune of someone else, we need to feel that their positive outcome is a result of our own actions. Third, we want other people to recognize how wonderful and helpful we are.

Still, given how many other people are stuck in traffic ahead of you and the high likelihood that they'll keep on letting cars merge into your lane, maybe you should start convincing yourself that real altruism consists of allowing good things to happen to strangers both directly and indirectly. And even when someone else gets the credit for it. Adapting this attitude won't be easy, but if you manage it, good things will follow.

Cars, Helping, Appreciation

ON IDLE WAITING

"It's O.K. I didn't marry you for your parking karma."

Dear Dan,

I noticed that when I drive around the block looking for parking I spend a lot of time very far away from my destination, which means that if I am lucky enough to find a parking space, I have to walk a long way in the cold (I live in Chicago). Because I hate the cold, I changed my strategy and now I just wait close to my destination until somebody leaves and I take their parking spot. It is hard for me to compare the two approaches directly, but waiting around seems to me to be as efficient if not better. The problem is that when I go places with some of my friends, they can't stand this strategy and I succumb to their pressure and keep on driving around looking for a parking space. My question is why do my friends find it so intolerable to wait for someone to leave?

—DANNY

The phenomenon you're encountering is aversion to idleness, and there's an interesting story about efficiency and waiting that is related to your question: a while ago there was an optimization engineer working for an airline, and he realized that some carousels were close to some gates, and others were close to other gates. He decided to optimize which carousel the luggage arrived on, such that the luggage from each flight would be delivered on the carousel closest to where the plane landed. Before this algorithm was created, travelers would get off the plane and walk such a long way that sometimes their luggage would be waiting for them on the carousel. After the new system was implemented, the carousel was much closer and people would walk just a short distance, find the carousel, and wait a bit for their luggage. People hated this new system because they were spending some of their time standing in one place waiting for their luggage (and to make things worse, maybe also wondering if their luggage was lost or not). This idleness was so unpleasant that people complained and the airline discarded this efficient algorithm.

My understanding is that the airline did not take the complete opposite approach and start unloading luggage at the farthest possible carousel to solve this specific customer service issue—but given their overall attitude to customer service, I suspect that they are actively working on this approach.

....

Dear Dan,

When I drive home at night, I have to look for a parking spot in my neighborhood. Should I stay in one place and wait for a parking spot to become available, or should I drive around in circles in search of a free space?

—IAN

I'm not sure there's an objectively correct answer, but here are a few things to consider. On the one hand, you never know when and where a parking spot will free up, but you know for sure that driving around wastes more fuel than staying put. This suggests that waiting in one spot is the right approach. On the other, if you idle in place, you might be waiting at a location where everyone has already parked for the night, and if you drive around, at least you get to spread your risk and hedge your bets. And this suggests that driving around is the right approach.

But you should also consider the psychology of waiting: Staying put and doing nothing is much more annoying than being active. When we just wait, time passes more slowly, and patience wears thin. Regardless of how much fuel they might save, a lot of people would go crazy if they had to just sit in their cars and wait. So between fuel economy and mood maintenance, the best thing to do might be to buy a fuel-efficient car and keep moving.

Cars, Time, Waiting

ON FORCING DECISIONS
WITH COINS

Dear Dan,

Do you have general advice for how to approach difficult decisions? I've been thinking about which car to get for a very, very long time, and I just can't decide.

—JOHN

Luckily the technology you need to solve this problem is already at your disposal. All you need is a coin. Assign each car to a side of the coin, and flip it high in the air. At this point, you can wait until the coin lands, and use this random device to make your choice—but my guess is that when the coin is in the air, you will realize which car is the one you really want.

The larger point is that once we have spent a substantial amount of time on a decision, and we still can't figure out which option is the best, it must mean that the overall value of the competing options is more or less the same. It is not that the options are identical, but that the difference in their overall quality is hard to distinguish. After all, if it were easy to tell which one is the best, we would have made the decision already.

Once we recognize that the decision is between options with similar overall value, we need to start looking at the opportunity cost of time. In an effort not to waste much more time on the decision, we need to force ourselves to make a decision. And this is what the coin method is all about. The coin goes up in the air, and at that point we are forced to realize how we want the coin to fall, face our preferences, and make up our minds.

Decisions, Luck, Emotions

ON TRASHY NORMS

"His bag of poop speaks well of him."

Dear Dan,

In the building where I live, we have a dedicated room for the large trash bins that serve the whole building. The problem is that some of the neighbors don't want to touch the dirty trash bins with their hands (and I understand this feeling of disgust), so they end up leaving their trash bags on the floor of the room, and someone else has to pick it up at some point (this selfishness I don't understand). Some of the neighbors in our building are politely asking the messy neighbors to put their garbage in the large trash bins, but to no avail. All kinds of threats have proven to be equally unsuccessful. What should we do?

—ARIEL

The problem in your building is not just about cleanliness. It is much more complex, and it has to do with the difficulty of changing social norms. What you have is a subculture in the building that does not consider leaving full trash bags on the floor to be embarrassing or shameful. And since this is the established norm for these individuals, it will take focused and deliberate effort to change this pattern of behavior.

In general, social norms are a powerful motivator and we are all susceptible to their influence in many areas of our lives. If you go to the trash room and see full trash bags lying around your judgment about right and wrong is affected, to some extent, by your own preexisting values and, to some extent, by the behavior of those around you. You say to yourself, "It seems that leaving the garbage bags on the floor is the standard practice in this particular place, which means that I can do the same and still feel good about myself." But if there is no trash around, you might tell yourself, "Leaving trash here is inappropriate and I shouldn't mess up this place." The important thing to remember about social norms is that when it comes to minor violations we need to relentlessly criticize the violators, because when the violations are repeated, the norm itself changes and with it the danger that everyone will be swept up in it.

And how can we create better social norms? I would summon a tenants meeting to discuss plans for the building. In the meeting I would try to create a new social understanding for what is proper neighborly behavior (taking care of the building, placing garbage in the right place, etc.). With the desired behaviors clearly defined, I would take another step and have everyone sign a pledge to follow these new guidelines. Finally, if you address these problems at a time that is close to a break point in the year (maybe New Year), I would use this symbolic time as an opportunity for change, and link the tenants meeting and the pledge to this new beginning. As long as you create a new social norm, the garbage will seem to deal with itself and the benefits might even extend beyond the trash room.

....

Dear Dan,

My partner and I live in a pretty 250-townhouse condo development and we have a problem with some people who don't clean up after their dogs. There's a fifty-dollar fine when an owner fails to clean up after their dog, but you have to know who the dog owner is, catch the person in the act, and report him to the condo corporation. This policy is not working. What can we do?

—RACHELLE

There are two approaches to consider in this situation: the positive approach of social norms and the negative approach of deterrence.

In terms of social norms, a great deal of research shows that what we do is less of a function of what's legal and not, and more in line with what we find socially acceptable. This means that if dog owners observe a lot of droppings around the condo area, they will find it perfectly acceptable to further contribute to this behavior. On the other hand, if they find the grounds to be pristine and poop-free, they will feel guilty leaving some doggy souvenirs behind. With this in mind, the first lesson from social norms is that violators are not only leaving an unacceptable mess behind them, but that they are also strengthening an undesirable social norm (more evidence for the popularity of this behavior), and therefore making it more likely that others will follow. The social norm perspective also means that you should work extra hard to establish a better social norm because once a more desirable social norm is set, the behavior can maintain itself.

In terms of deterrence, I think you should try something more exotic than a fifty-dollar fine. I suspect that right now some dog owners see the setup as a "game" where they leave the droppings, while the other neighbors and managers try to catch them in the act and fine them. Assuming that this is the case, I would try to alter the nature of the game such that it unites the community of dog owners. For example, what if the condo management put a fixed amount of money in a community fund that was designed to pay for a droppings cleaner, as needed, and whatever money was left at the end of each month was used for a get-together for all the dog owners and their dogs? If a lot of money remained in the doggy-bank, the party would include food, drinks, and doggy treats; if no money was left, the party would serve water. This way, failing to clean up after their dogs would negatively impact the dog owners and their community. By increasing the personal and social costs of leaving dog poop behind, this kind of mechanism should get people to be more thoughtful, and make the grounds poop-free.

Social Norms, Other People, Coordination

ON MAKING SMOKING FEEL DANGEROUS

Dear Dan,

What's the best way to get people to stop smoking?

—MYRON

The problem with smoking is that its effects are cumulative and delayed, so we don't feel their danger. Imagine what would happen if we forced cigarette companies to install a small explosive device in one out of every million cigarettes—not big enough to kill anyone but powerful enough to create a bit of damage. My guess is that this type of immediate danger would make people stop smoking. And until we find a way to implement this approach, maybe we can get people to at least start thinking about smoking this way.

Health, Self-Control, Habits

ON ADVENTURES
AS INVESTMENTS

Dear Dan,

I will graduate from college next year, and I really want to teach English in Spain afterward. There's still a market for this, despite the Spanish economy, but I'm wondering if I should do it or not. It would give me an unforgettable experience—one I don't think I'll be able to have when I'm settled down with a job, a husband and kids. But, at the same time, it could delay the start of my career, which I want to have on track before I settle down. Is the experience worth delaying the start of my "real life"?

—GABRIELLA

The question is not about when to start your career. Your career has been on its way for a while now. What you are really asking is what is the best next step on this path.

When I graduated, I asked Ziv Carmon—one of my academic advisors—which university I should aim for as my first academic job. His answer was that I should go to the place where, five years down the road, I would be most likely to emerge as a very different person. He then explained that life is about learning and improving, and that I should take advantage of my relative flexibility (no wife and kids at the time) and invest the next few years in my own development and growth. I took his advice seriously and accepted a position in a place that was not the best fit with my existing knowledge base. As a consequence, over the next few years I learned a lot of new things, changed my interests substantially, and became a much better researcher and teacher. I even think that I became a slightly better person (but maybe that part was just the aging process).

Since then, I have been a fan of thinking of the early years of life as an opportunity to collect lessons and experiences so that we are better equipped for the long and unpredictable road ahead. Of course, it is hard to predict what life lessons and experiences will be most useful for us in the future, but if you collect many experiences and skills, there is a good chance that some of them will become highly useful. Maybe you can think of this time in Spain as gambling with your time now for a future benefit. And since the seeds you sow now can yield fruit over many, many years, I would go for it.

Workplace, Education, Long-Term Thinking

ON THE QUALITY AND NOT THE QUANTITY OF IRRATIONALITY

Dear Dan,

I am convinced that some of our decisions are irrational, but what's the proportion of irrational decisions?

—JULIANNE

The right question, I think, isn't what's the proportion of our irrational decisions—it is about their impact on our life and well-being. Think about something like texting and driving. Perhaps we do it *only* 3 percent of the time, but each of these instances could kill us and other people. So what we really need to ask ourselves isn't about the proportion of our irrational behaviors but about the extent to which such irrational behaviors can harm us, those around us, and society in general. When we think about our behaviors this way, it seems to me that the impact of our collective irrational decisions on our lives is very very large.

Decisions, Mistakes, Regret

ON "HELPING"
PEOPLE RETIRE

Dear Dan,

What is the best way to make sure Americans have adequate funds for retirement?

—BEN

There are basically two ways to help people get sufficient money to fund their entire retirement. The first is to get people to save more money, and to start saving at a younger age. The second approach is to get people to die at a younger age. The easier approach, by far, is getting people to die younger. And how might we achieve this? By allowing citizens to smoke. By subsidizing sugary and fatty foods. By limiting access to preventive health care etc. When we think about retirement savings in these terms, it seems that we're already doing the most we can on this front.

Long-Term Thinking, Health, Decisions

ON THE MORALITY OF CORRECTING MISTAKES

Dear Dan,

I just paid for yoga classes for the next six months, but the studio mistakenly credited me as if I paid for a year. In the past they have made many billing errors, but all of these were in their favor. Should I correct the mistake or just see it as the universe making things more even?

—A RANDOM FAN

There is no doubt in my mind that it is not a mistake. It is a simple case of the world restoring karma—my only question is, why did it take such a long time?

Honesty, Luck, Value

ON WHO WE ARE
AND WHO WE WANT TO BE

"Well, instead of discussing the book we could discuss why
none of us had time to read it."

Dear Dan,

I have been on vacation for the last few days, and while reading your book on dishonesty, I have been wondering whether people behave more or less honestly while on vacation.

—JULIE

This is an interesting question, but sadly I don't have any data on this topic. Nevertheless, here are a few possibilities to consider: One reason why we might be more honest is that while on vacation we might put aside some of our concerns about money. To the extent that the motivation to be dishonest is based on financial gains, and to the extent that we are less concerned with money, we might be more honest. A second reason why we might be more honest is that on vacation we are often in a good mood, and there is some evidence to suggest that when we are in a good mood we are willing to do a lot to keep ourselves in that state, suggesting that we are less likely to take risks that might spoil our good mood.

On the other hand, there are also some reasons to suspect that while on vacation people are likely to be less honest. One reason is that vacation takes place in a new and unfamiliar context, which means that immoral behavior in this atypical context will not have the same negative implications for how we perceive ourselves. A second reason for more dishonesty while on vacation is that the rules on vacation might seem less clear and easier to bend: What are the regulations for jaywalking? How much should we tip in Portugal? Is it OK to take the towels from this hotel in Turkey? This sort of wishful blindness can make it easier for us to misbehave while still thinking of ourselves as wonderful, honest people.

On balance, then, are we more or less honest when we are on vacation? I suspect that we are less honest. But I would love to be proven wrong.

Honesty, Emotions, Self-Image

ON THE VALUE
OF SPLITTING CHECKS

"I like to sit facing the room to see if anyone
seated after us gets served before us."

Dear Dan,

When going to dinner with friends, what is the best way to split the bill?

—WILLIAM

This is certainly an important question as it involves the delicate fabric of friendship, social justice, and the optimal design of experiences.

There are basically three ways to split the bill. The first is for everyone to pay for what they've had, the second is to split the bill evenly, and the third is for one person to pay for everyone, and to alternate who pays over time. I like the alternating approach the most, the equal pay second, and the exact payment the least. Here is why:

If you take the pay-just-for-myself approach, every person must become a part-time accountant, identify their items on the receipt, make a note of the prices, and tally up their bill. To add insult to injury, this annoying accounting process comes at the end of the evening, and because the end of an experience plays a large role in how we remember the experience as a whole, it can cast a dark shadow on the whole evening.

Another approach is to share the bill equally, which works well when people eat (more or less) the same amount. But, again, contrast what it would feel like to end an evening remembering how delicious the crème brûlée was versus ending an evening thinking how annoying it was that Suzie ate so much more of the main dish, but paid the same amount.

The final approach (my favorite) is to have one person pay for everyone and to alternate the designated payer with each meal. If you go out to eat with the same group relatively regularly, it ends up being a much better solution. Why? First, getting a free meal is a special feeling, and this approach maximizes the number of people who feel that they got a free meal. Second, while the person paying for everyone ends up feeling worse about paying a large amount, the elation of the other people who got the free meal more than compensates for this increased negative feeling. In economic terms this means that the social welfare is higher. And third, the person buying may even benefit from the joy of giving. Let's think about an example with two friends, Jaden and Luca, who are going out to their favorite Middle Eastern restaurant. If they were to divide the cost of the meal evenly, each would feel, say, 10 units of misery. But if Jaden pays for both of them, Luca would have zero units of misery and the joy of a free meal. And because of diminishing sensitivity as the amount of payment increases, Jaden would suffer fewer than 20 units of misery—maybe 15 units. On top of that, Jaden might even get a boost in happiness from getting to buy his dear friend a meal.

Taking all of these elements into account you can see why it is best for one person to pay for everyone and to alternate who pays over time. And what if you do not go out with the same exact group every time? Even in these cases I think it is worth it, because the increase in overall joy from using this approach is so large that it easily compensates for the occasional financial loss.

Friends, Food and Drinks, Spending, Emotions

ON STAPLERS
AND QUARTERS

"When you lie about yourself, is it to appear closer to or
farther away from the middle of the bell curve?"

Dear Dan,

I was talking with a friend about one of your experiments on dishonesty in which people felt free to steal sodas and cookies from the "break room" but not the equivalent amount of cash. My friend said that in his workplace items such as staplers, tape dispensers, and so on used to be constantly taken from his desk. He then glued a quarter onto each piece, and no one has taken anything with a coin on it for five years. Does this fit with your findings?

—TONY

This is exactly the point. It turns out that we can rationalize lots of our bad behaviors, and the more distant they are from cash the simpler it is for us to rationalize them. What your friend has done by sticking money to the items is to make it clear that borrowing the office supplies without returning them is not just about the office supplies, it is also about stealing cash. And with this reframing he made the action more morally questionable in the minds of the potential thieves.

I love the application of this principle to the office environment. Now, if we could only glue quarters to stock certificates and other financial products, maybe the world would be a better place.

Honesty, Money, Workplace

ON TAKING TIME
FOR EXERCISE

"I'm going out to get some endorphins."

Dear Dan,

There are many people in my office who have a hard time focusing for even twenty minutes on their jobs. Nevertheless, they seem perfectly capable of exercising for long stretches, and amazingly they are quite persistent in their ability to focus and sustain long periods of physical activity. Can you explain this contradiction?

—MICHAEL

This might actually not be a contradiction but rather, as I learned recently, two faces of the same mechanism. A few weeks ago, I flew to California for some meetings. I left home at 4:30 a.m. and got to San Francisco at 10 a.m. I had my set of meetings, and by 5 p.m. was exhausted. I had a lot of work-related tasks that I was behind on, and I was determined to get at least some of them done. The problem was that I felt devoid of any energy. So I went for a run.

Ordinarily, I try to run regularly, maybe once every five or ten years. But this run was fantastic and it changed my view of running. I ran a bit, walked a bit, listened to music along the way. It was challenging, and I was quickly out of breath, but in no way was it even close to the mental exhaustion of doing the things I was supposed to work on. I was basically shirking and feeling good about it.

Here is my new understanding: I think that people who either don't enjoy what they're doing for work or don't have the mental stamina to focus on the complex tasks that they need to attend to are more likely to take long breaks for exercise. Imagine that some of your coworkers took a two-hour break to read a book or watch a movie. They would certainly be seen as selfish slackers who are wasting time and not contributing anything to their organization or society. But because our social rules tell us that exercising is good for our health, it is a perfectly reasonable excuse to escape work, feel good about it, and get respect from the people who are left behind to cover for those who are exercising.

Now that I have discovered this approach to taking time for myself without feeling guilty about it, there is no question in my mind that I am going to go for runs more often.

Workplace, Exercise, Procrastination

ON MEMORY

"I'm not losing my memory. I'm living in the now."

Dear Dan,

How can I enjoy life more? Every year, time seems to go by faster; months rush by, and years just seem to disappear. Is there a reason for this, or is the memory of time passing more slowly when we were children just an illusion?

—GAL

Time does go by or, more accurately, it feels as if time goes by more quickly the older we get. In the first few years of our lives, everything we sense or do is brand-new, and a lot of our experiences are unique—so they leave a strong impression, and remain firmly grounded in our memories. But as the years go by, we encounter fewer and fewer new experiences. One reason for this is that by the time we reach maturity we have already encountered and accomplished a lot. Another, much less happy reason is that over time we become slaves to our daily routines and we try fewer and fewer new experiences.

To see if this is indeed the case for you, just try to remember what happened to you every day during the last week. Chances are that nothing extraordinary happened, and that you are hard-pressed to recall the specific things you did on Monday, Tuesday, Wednesday . . .

Given the importance that remembering our experiences has on our life-satisfaction and happiness, what can we do about this worrisome trend? Maybe we need a memory/experience app that will encourage us to try new experiences, point out things we've never done, recommend dishes we've never tasted, and suggest places we've never been. Such an app could make our lives more varied, prod us to try new things, slow down the passage of time, and increase our happiness. And until such an app arrives, how about trying to do at least one new thing every week?

Aging, Memory, Experiences

ON BOOKS
AND AUDIOBOOKS

"I got tired of *Moby-Dick* taunting me from my bookshelf,
so I put it on my Kindle and haven't thought of it since."

Dear Dan,

From time to time, people around me discuss a book they have read recently. While I know the book well, and I want to participate in the conversation, I hesitate because I listened to the audio version of the book. My first question is, why am I embarrassed to say that I listened to the book? My second question is, what can I do about it?

—PAULA

We learn how to listen and comprehend at a young age and therefore we don't really remember how difficult listening and comprehending spoken language was at some point for us. On the other hand, we learn how to read and write at a later age and we remember the difficulty of our early struggles with reading and writing all too well. Because of this difference, people associate greater difficulty with reading than listening, and as a consequence, we take greater pride in reading than listening.

My first suggestion is that you remember that this isn't necessarily the case and that reading is not necessarily more difficult than listening. In fact the order of difficulty might be reversed. When I received your question I went ahead and purchased an audiobook and I listened to it on a long flight. For what it is worth, I found it harder to focus while listening to a book than reading a book. (I should also point out that the book I listened to was *Galápagos* by Kurt Vonnegut, so this might explain some of the difference.)

A second suggestion is that you find a different word to describe your experience. For example, for books you loved, maybe you can say: "I inhaled that book." For more difficult books, maybe you can say: "I struggled with it."

If these approaches don't work for you, perhaps it is time to expand the meaning of the word *read*. Maybe we should acknowledge that today there are many ways to get information—audiobooks being one of them. This might seem dishonest, but you might be able to lead a naming revolution and help lots of people who listen to audiobooks feel more comfortable with what they're doing.

Technology, Entertainment, Language

ON SOULS AND PASCAL'S WAGER

Dear Dan,

Out at a bar recently, I met someone who told me that he did not believe that people have a soul. I immediately asked him if he would sell me his. We ended up agreeing on a price of twenty dollars. I paid up, and he wrote a note on a napkin giving me his soul.

Now, I don't believe in an afterlife, but I also can't help but believe that there is an exceedingly small chance that a soul could have a much higher (maybe even infinite) value. So twenty dollars seemed a reasonable hedge. Did I pay too much? Or did I get a good deal?

—CAREY

Well haggled. Your logic here is reminiscent of what is known as Pascal's Wager, named after the philosopher who figured that if there was even a small probability that God and heaven exist, and assuming that the afterlife is infinite, the smart move is to live your life as if God and heaven exist (because you would multiply the small probability by infinity, and the outcome would be infinity).

In terms of the cost, I suspect that you got a very good deal for three other reasons. First, discussing this trade must have been far more interesting than the usual bar chitchat, so if you value the quality of your time, the twenty dollars was a good investment even if it turns out that souls don't exist. Second, you now have a great story to reflect on for a long time, which is also worth something, and most likely more than twenty dollars. And finally, you are now the proud owner of a soul. But if all of these reasons don't convince you, send me the soul, and I'll happily buy it from you.

Regret, Value, Religion

ON SHOWING
OFF THE PRICE

"Oh, great! Here comes Valerie to raise the bar."

Dear Dan,

I bought two different bottles of wine at a wine store that was running a "Buy one, get another for five cents" deal. The first one cost me $16.99, and I got the second for five cents.

Tomorrow I am invited to a close friend's house for dinner, and I'm going to take one of the bottles, but which one should I take? Should I take the $16.99 one or the five-cents one, and should I tell him about the cost?

—RAGS

We've known for a long time that the more expensive a wine is, the more we enjoy it. However, the particularly interesting thing is that this correlation exists only when we know the price. When the wine tasting is blind, and we don't know the price, there is virtually no correlation between the cost of the wine and how good we think it is (experts have a positive correlation even in blind tasting, but there are very few of these real experts, and even for them the correlation is very small).

Taking this into account, the first question you should ask yourself is whether to tell your friends about the cost of the wine or not. If you don't tell them, then there's no problem—just take the cheap one. By knowing how little you paid for the wine, you will enjoy it less, but everyone else will be just fine and maybe you should just drink a different wine. On the other hand, if you decide to share the price, I would suggest bringing the $16.99 bottle, and maybe even pad the cost a bit by including the expense of driving to the wine store and the value of your time. This way your friends are sure to enjoy your amazing wine.

Food and Drinks, Value, Expectations

ON TOPICS
AND TEACHERS

"I need you to line up by attention span."

Dear Dan,

I am a student in middle school, and there is one subject in school that I really love, and one subject that I deeply dislike. There is also one teacher I really love and one teacher I am not very excited with. Here is the question: would I be better off if the teacher I love taught the topic I love, and the duller teacher taught the topic I dislike? Or would I be better off if the teacher I love taught the topic I dislike, and the duller teacher taught the topic I love?

—AMIT

What you are really asking about is the accumulation of pleasure and pain. Let's call the approach of combining the two good components and combining the two negative components the extreme approach (since one class is going to be great and the other one terrible) and let's call the approach of mixing a good and a bad component the average approach (since both classes are going to have some good and some bad).

If you believe that the nature of combining experiences is asymmetric, such that every additional bit of good experience makes it better, but that once something is negative it doesn't matter how negative it is—then you should go with the extreme approach. This way you would have at least one wonderful class to look forward to. It is also true that this approach would leave you with one really bad class, but if you believe that once a class is going to be bad, it doesn't really matter how bad it is, then the extreme approach is the one for you.

A very different set of beliefs about combining experiences is about the ability to tolerate extreme negative experiences. If you believe that a class with both a bad teacher and a bad topic is going to be too much to bear, that the combined pain will push you over the edge, and that this level of misery will darken your entire semester, then the average approach is the one for you.

From my perspective I am first and foremost delighted that you like some of your teachers and topics. On top of that it is very important that you don't stop thinking of school as a joyful, exciting place. Given that you are going to spend many more years in formal education, and then the rest of your life learning in many other ways, it is important that you love learning. Given this, I think that the average approach would be better for you. I suspect that having a class with both a bad teacher and a hated topic will be too much for you to handle, and that it might make it hard for you to love school and continue learning.

One final point: I also suspect that if you had a class with the teacher you love and the topic you don't, you will learn to focus on the teacher and pay less attention to the topic, while in the class with the teacher you dislike and the topic you love, you will learn to focus on the material and pay less attention to the teacher. This means that by shifting attention to the part of the class you enjoy, you might get more value and less misery from each of the two classes.

I wish you many years of exciting and pleasurable learning.

Education, Experiences, Motivation

ON (THE LACK OF) SELF-CONTROL

"When portions are this huge, I eat half now
and the rest in a few minutes."

Dear Dan,

Whenever I stay up late, I wind up raiding the fridge—and ruining my diets one after the other. During the day, I manage to resist the temptation, but at night, my self-control seems to stop working. What should I do?

—MENI

What you describe is a well-known phenomenon called "depletion." All day long, we face small temptations and do our best to resist them. We maintain control over ourselves and the temptations around us in an effort to be productive, responsible people and stop ourselves from giving in to our urges to shop, procrastinate, watch that latest cat video on YouTube, and so forth.

But our ability to resist urges is like a muscle: The more we use it, the more tired we become—until at some point, which is very likely to be at night, our willpower simply becomes too weak to stop us from giving in to temptation. This is one reason the temptation industry—bars, strip clubs, etc.—operates mostly at night: After we have been resisting temptations all day, we are depleted and ready to fail, and these temptation-institutions are ready to profit from our failure.

One way to overcome this depletion problem is based on the story of Odysseus and the sirens. In this story Odysseus told his sailors to tie him to the mast and to not untie the ropes, under any circumstances, until they sailed past the sirens. This way Odysseus couldn't act on his temptations, and jump into the water and swim toward the sirens' seductive calls. The modern equivalent of this tactic? Keep all tempting food out of your house. You can hope that your future self will be able to resist temptation, buy the chocolate cake, and eat just a sliver of it every other day. But the safer bet is to realize how easy it is for us to fail, how much easier it is for us to fail toward the end of the day, and simply not keep any chocolate cake in the fridge in the first place.

Dieting, Self-Control, Food and Drinks

ON THREE
BUILDING BLOCKS
OF A BALANCING ACT

Dear Dan,
   Why do consultants always break problems and solutions into three?

—ALICE

When consultants give answers, they often try to strike a delicate balance between making the answer simple and making it feel complete. I suspect that offering three things to consider strikes this sweet spot.

Communication, Appreciation, Decisions

ON WASTING TIME
DECIDING

Dear Dan,

Often when I meet with my regular group of closest friends, the discussion turns out to be something like this: "Where do you want to go?" "Not sure." "Where do you want to go?" "Not sure." These discussions are frustrating, uncomfortable, and a waste of time. Any advice on how to move them forward and get to a decision faster?

—MATTHEW

When someone asks "What do you want to do tonight?" what they are implicitly saying is: "What is the most exciting thing we can do tonight, given all the options and all the people involved?"

The problem is that figuring out the absolute best solution (the optimal solution) is very difficult. First, we need to bring to mind all the possible alternatives; next we need to figure out our preferences and the preferences of all the people in the group. Then we have to find the one activity that will maximize this set of constraints and preferences.

The basic problem is that, in your search for the optimal activity, you are not taking the cost of time into account. You waste your precious time together asking "What do you want to do?" which is probably the worst way to spend your time.

To overcome this problem, I would set a rule that limits the amount of time you are allowed to spend searching for a solution, and I would choose, in advance, a default activity in case you fail to come up with a better option. For example, take an acceptable good activity (going to drink at X, playing basketball at Y) and announce to your friends that, unless someone else comes up with a better alternative, in ten minutes you are all heading out to X or Y.

I would also set up a timer on your phone to make it clear that you mean business and to make sure that the time limit is honored. Once the buzzer sounds, just start heading out to X or Y, asking everyone to come with you and telling the people who do not join you immediately that you will meet them there. After repeating this tactic a few times, your friends will get used to it and you should experience an end to this wasteful habit.

Friends, Decisions, Coordination, Time

ON BUFFET ROI

"I set a goal, I met it, I proved that I could meet it,
and now the hell with it."

Dear Dan,

How should I maximize my return on investment at an all-you-can-eat buffet? Should I go for dessert first and then hit the entrees? Or should I start with the salads and then pick only healthy foods from the main courses?

—SYED

While I appreciate this return-on-investment, or ROI, mindset, in food, as in all other areas of life we must focus on the right type of returns. In particular your mistake here is that you seem to focus on the short-term returns, and not the long-term ones. If you go into a buffet trying to maximize your short-term ROI, you might gulp down more food but then you'll have to deal with the long-term effects of your actions. Maybe you will end up spending extra hours at the gym or maybe you will add even a little more weight to the few extra pounds that so many of us are packing. Whatever approach you pick to deal with the extra food intake, there will be some consequences to your short-term optimization.

A different type of mistake that some of us make when we go to buffets (and this mistake also shows up in other areas of life) is to focus on maximizing the cost of the food to the buffet's operators. The basic heuristic is that we somehow feel that we are getting a better deal when we eat food that is more expensive. But, of course our goal in life (as in buffets) is to maximize our enjoyment, and not the cost to someone else. If we enjoy good bread and cheese more than caviar, maybe we should try a bit of caviar from time to time just to be sure that we indeed don't like it, but we should focus on the food that brings us higher enjoyment.

And to your question: What I recommend is to mostly stick to a balanced and healthy diet. But since many buffets boast a large assortment of novel dishes, and in the spirit of the idiom that variety is the spice of life, I would make some exceptions and sample one or two delicacies that I've never tried before. Just for the experience.

Food and Drinks, Long-Term Thinking, Experiences

ON ASKING THE RIGHT QUESTIONS

"I'm old enough to recognize a lecture
disguised as grace."

Dear Dan,

My daughter recently started dating a lazy, dumb guy. How can I tell her gently that he is wrong for her without sounding preachy, causing her to ignore me completely, or even worse, go against my advice on purpose?

—A CONCERNED MOTHER

Without knowing the real quality of your daughter's boyfriend, I should point out that what you are experiencing might be the common reaction of parents around the world when their (perfect) daughters bring home their (far from perfect) boyfriend. I don't even want to imagine my own reaction the first time my daughter (Neta, who is 9 years old) introduces me to her potential romantic partner . . .

But let's assume for the sake of argument that you are, in fact, correct, and that your daughter's new boyfriend really is dumb, lazy, and up to no good. Even with these starting qualities you shouldn't tell your daughter your true opinion. Instead you should ask her some questions—leading questions, of course. Under normal circumstances we tend to avoid asking ourselves difficult questions, but if someone else asks us questions there is a chance that these questions will be planted in our minds and become part of our inner monologue. For example, you might ask, "How do you and your boyfriend get along? Do you ever fight? What are the ten things you love most about him? What do you like least about him?" The answers to these questions are likely to be: Fine; A few times a week; I can think of three, but I can't think of ten; His selfishness. As a result, your daughter might start paying attention to every fight, to the few things she really loves about him, and to his selfishness.

I admit that this approach is a bit manipulative, but I hope it will get her to think in depth about her relationship with this boy and maybe she will reach the same conclusions you have.

Family, Relationships, Opinions

ON DOUGHNUTS
AND THE LOCUS
OF FREE WILL

Dear Dan,

If people make decisions in a way that largely depends on their environment, does that mean that there is no free will?

—MATT

Yes and no. Imagine that every day, I stopped by your office first thing in the morning and covered your desk with fresh doughnuts. What are the odds that by the end of the year you will not weigh substantially more? My guess is close to zero. Once the environment is set, our behaviors largely follow. But the good news is that we don't have to be tempted by doughnuts every day. We can keep the doughnut peddlers out of our office and, more generally, we can set up our environments in ways that reduce our potential for failure.

That's where our free will resides—in our ability to design our environments in a way that is more compatible with our strengths and, even more important, helps us overcome our weaknesses.

Decisions, Mistakes, Long-Term Thinking

ON THE
MOST OPTIMISTIC
DAY OF THE YEAR

Dear Dan,

Do you believe in New Year's resolutions?

—JANET

Yes. Very much. Every year for about a week: for about five days before the end of the year and for about two days after New Year's Day.

Habits, Self-Control, Wishful Thinking

ON EMOTIONAL INVESTING IN THE STOCK MARKET

"How can a fund that's losing us money
call itself socially responsible?"

Dear Dan,

How can I control myself when I feel the irresistible need to break my own rules about how to invest in the stock market?

—GANAPATHY

You are asking, I suspect, about what we call the "hot-cold-empathy gap," where we say to ourselves: "The level of risk that I want to take is bound on one side by gains of up to 15 percent and on the other by losses up to 10 percent." But then we lose 5 percent of our money; we panic and sell everything.

When we look at such cases, we usually think that the colder, more rational voice in our head (the one that set up the initial risk level and portfolio choice) is the correct one and the voice that panics while reacting to short-term market fluctuations is the one causing us to stray.

From this perspective, you can think about two types of solutions: The first is to get the "rational" side of yourself to set up your investment in such a way that it will be hard for your emotional self to undo it in the heat of the moment. For example, you can ask your financial advisor to prevent you from making any changes unless you have slept on the decision for seventy-two hours. Or you can set up your investments so that both you and your significant other have to sign some document in order to make any changes to your investments. Alternatively, you can try to not even awake your emotional self, perhaps by not looking at your portfolio very often or by asking your financial advisor to alert you only if your portfolio has lost more than the amount that you indicated you are willing to lose.

Whatever you do, I think it's clear that the freedom to do whatever we want and change our minds at any point is the shortest path to bad decisions. While limiting our freedom often goes against our ideology, sometimes limiting our ability to make decisions is the best way to guarantee that we will stay on the long-term path we intended to take.

Loss Aversion, Stock Market, Emotions

ON COMMUTING
AND ADAPTATION

"I'm sorry, Jim. I love you, but I hate Vermont."

Dear Dan,

We recently got married and are having a hard time deciding where to live. Should we live in the city, close to where we work? Or would we be better off finding someplace cheaper, greener, and farther away from the city?

—A COUPLE FROM THE CENTER

Your decision should take a few things into account. First, most of us can get used to most things in life: different-size houses, a neighborhood that is lush or drab. And we adapt to most of these changes faster than we expect. My own personal example of this is that many years ago I suffered a serious injury that changed my life dramatically in almost every aspect. But over time I got used to these changes, and now my life is much better than I could have imagined when I was originally injured.

While we adapt to many things rather successfully, there are some things that we don't adapt to, or at least not that easily. One of these, sadly, is commuting—that annoying daily trip from the small neighborhood where we live to our place of work in the big city.

Beyond the fact that we have a hard time adapting to commuting, the reason that we don't get used to it provides an important insight into the nature of the adaptation process. If we knew that we could leave home each day at 7:30 a.m. and arrive at work at 8:55 a.m., commuting would be predictable and expected, and we would quickly adapt to it. But because we never know what is in store for us in terms of traffic and bottlenecks, we never know when we might arrive at work. This uncertainty makes it difficult to get used to commuting, and it makes us start each day with a constant worry of whether we will get to work on time or not.

This is why I suggest that you take distance from work into account as a significant factor in deciding where to live. It will likely play a larger role in the quality of your life than you expect.

Commuting, Happiness, Adaptation

ON RIDING YOUR
DRYER TO TUCSON

Dear Dan,

I'm shopping for a flight for a personal trip that will take place in a few months, and I keep running into the same problem: "Current me" wants to pinch pennies by choosing an overnight flight with several legs and an inconvenient airport that would require me to drive a few hours out of my way. "Future me"—the one that will actually pick up the rental car at 11 p.m. and drive two hours from Phoenix to Tucson the night before my friend's wedding—will resent that I was saving the extra hundred dollars instead of trying to make an already expensive trip more pleasant. Travel-booking websites are getting better and better at predicting what will happen to flight prices, but I don't seem to have gotten any better at predicting my own preferences.

How can I best determine whether these savings will feel worth it to me in the future? Or, failing that, how can I console myself when I'm pulling into a Tucson motel parking lot at 1 a.m.?

—RUTH

Your framing of the problem is spot-on. In your current "cold" state, you pay attention to the price, which is clear and vivid and easy for you to focus on and think about. When you will actually be on the trip, you will be exhausted and in need of sleep (a "hot" state), which will be painfully apparent to you at that point. But all of this is not as vivid to your current self as you sit comfortably and well rested at your computer, comparing the different travel options.

This, by the way, is a common problem that arises every time we make decisions in one emotional state about a consumption experience that will take place under a different state of mind and a different emotional state.

Here is what I recommend. In order to make a better decision, tonight at 9 p.m. put in some laundry and spend the next two hours sitting on the washer and dryer. This is to simulate the fun of flight. If you want to really go all out, supply yourself with a package of peanuts and a ginger ale. When you "land" at 11 p.m., look around for some missing socks (to simulate looking for your luggage) and then, properly primed to better understand how the actual experience will feel, log into the travel website and see what is more important to you: saving a few bucks or getting to bed sooner.

And to make the simulation even more effective, try to imagine how you will look in the wedding pictures after a long night of uncomfortable travel.

Good luck with your decision and "mazel tov" to your friend.

Travel, Predictions, Emotions, Decisions

ON PROMOTIONS
AND THE ILLUSION
OF PROGRESS

"If you don't want to feel inferior to your classmates,
you shouldn't have gone to such a good school."

Dear Dan,

I work in high tech but can't seem to get ahead. A good friend of mine is in the police force, and he gets promoted all the time. He claims that it has nothing to do with his skills, but with the lower quality of the people working there. Would it be better to choose a line of work where everybody else is mediocre and I'm the best instead of picking a high-profile workplace where I continuously feel unpromotable?

—DAVID

Your general question has to do with the joy people derive from feeling that they are advancing and developing in their careers. This feeling of progress is very important to our well-being and it provides gratification, self-esteem, and recognition from our peers.

Widespread recognition of the need for progress explains why so many companies have invented titles and intermediate positions for management types (officer, executive, chief, vice president, senior vice president, deputy CEO, etc.). In a gamification-like approach companies want their employees to feel that they are making progress and moving ahead even when these steps are not very meaningful.

Historically, this trend only affected those who were on the management path. Engineers remained engineers, even when their salaries increased and their responsibilities expanded. But over the years, companies have embraced the strategy of creating new titles across many departments within their organization. Now most companies, across all positions, have a list of titles that give all employees the feeling that we are moving ahead on this title treadmill (and academia is certainly a leader in this process).

Given the elegance, simplicity, and ease of feeling a sense of progress—and before you quit your job for one where your coworkers aren't going to be as good—see if you can receive, or even create, a promotion with a new title. Speak to your boss. Try to increase your responsibilities. Suggest a new title on your business card. With these changes you will certainly achieve your well-deserved feeling of accomplishment.

And if this approach doesn't work, talk to more people, maybe even find some new friends who are not doing as well at their jobs as your policeman friend. You may find that you are extremely successful compared to some of them, and if you stick with these less successful friends you will feel better in comparison.

Workplace, Comparison, Goals

ON DISTANCE FROM EMOTION AND CARING

Dear Dan,

I am writing to you from a train in Germany, sitting on the floor. The train is crowded, and all the seats are taken. However, there is a special class of "comfort customers" who are allowed to make those already seated give up their seats. This status is given to those (like me) who travel a lot on the train. Obviously, it would be nice to get a seat and, according to the rules, I deserve one. But I can't see myself asking one of the "non–comfort customers" to give up their seat for me. Why is this so difficult for me?

—FREDERICK

Your question has to do with what is called the "identifiable victim effect." The basic idea is that when we see a person suffering up close, our hearts go out to them, we care about them, and if we can, we help. But when the problem is very large or far away, or when we don't see the person who is suffering, we don't care to the same degree and we don't help.

In your case, I suspect that if ten minutes before you boarded the train the train conductor picked a random passenger to clear their seat for you, the situation would place a large psychological distance between you and the victim and, as a consequence, you would be perfectly able to enjoy the seat.

And what if the person giving up their seat was more identifiable? For example, what if the conductor pointed the seat victim out to you as you boarded the train? Or even worse, what if the process of evacuating the person from their seat was done in front of you? The worst of course would be for you to pick the person, inform them that your comfort demands that they leave their seat, and watch their reaction to the news.

What's the lesson here? Direct contact with other people makes the impact of our actions more vivid. It causes us to feel, empathize, and act with more care and compassion. And the big question is how to get our politicians, bankers, CEOs, and everyone else who has remote impact on our lives feel the consequences of their decisions and actions.

Travel, Other People, Emotions

ON PREDICTING
HAPPINESS

"I'm afraid of the person I might become if
I ever left New York."

Dear Dan,

Should I quit my job? I'm generally unhappy with it, but I've been with the company for eight years, and there are several practical and financial reasons to stay: I make a good salary, including stock options and grants; I get several weeks of vacation each year; and I have a pension. In addition, there is a lot of uncertainty with starting over in a new job, and there is no telling whether I would be any happier in a new place. Any advice on whether I should stick with what I know or if I should look for fulfillment in a new place?

—KP

Your question is really about the roots of your unhappiness. Is it caused by the job or by you? If your unhappiness is based on your particular job, then switching is a good path to a better future. On the other hand, if the cause for your misery is you, then switching is going to be of no use, because, as the saying goes: "no matter where you go, there you are," and you will still be there to make your new job just as miserable.

So, which one is it? It is hard to tell, and in particular it is hard to tell because you have been in the same job for a long time, and you have no evidence for how you would feel under different circumstances and in a different job (by the way, you can think about analogous questions in romantic relationships).

With this difficulty in mind, I would suggest that you take your next vacation (let's say three weeks) and use the time to volunteer at the kind of a company you are considering moving to. Note that I am recommending a rather long trial period because you need some time to get a better sense for how this other job would feel once the initial newness wears off. Of course, a few weeks as a volunteer would not give you the full sense of what it would feel like working at that company for a long time, but it would give you a much better sense of the root cause of your current joyless state. At the end of this trial period you should have a better idea if the cause for your unhappiness is your current job or you.

And one more thing: If you don't think that spending three weeks of your vacation trying to figure out if you should stay at the same job or move, you are probably not really that unhappy and you should stay where you are and stop complaining.

....

Dear Dan,

Seventeen years ago when my son was eighteen he moved to New York to attend The Cooper Union as an art student. He is now thirty-five and afraid to leave NYC. He doesn't like living in NYC and says he would love to move west, to be with us and to develop his art (and hopefully a career) in photography. But the people around him are giving him very different advice. It seems that people in New York believe that it is the only place in the world to live and to get a job. Is there any advice or constructive approach you can offer to make it clear to him that he is stuck and should move west?

—BARBARA

First, it's delightful that you want your son to move closer to you rather than stay on the other coast, and I am sure that he feels the same.

In terms of his moving versus not moving, I suspect your son is suffering from a combination of three decision biases. The first is the endowment effect, which has to do with our tendency to use our current situation as a reference point, and view any other alternative as a negative change from where we are now. In your son's case, moving from New York City to the West

Coast has some advantages (weather, his parents, etc.) and some disadvantages (lower density, fewer art galleries, etc.), and the endowment effect suggests that he is focusing to a larger degree on the things he would give up, and not paying sufficient attention to the things that he would gain if he ever moved to the West Coast.

The second decision bias your son is most likely suffering from is the status quo bias, which means that we feel very differently about a decision to stay in a situation, compared with a decision to change our situation. I once heard an air force commander tell his pilots that every second, they are making a decision to change course or to stay their course, and that they should always think about their actions as active choices. The problem is that very few of us think about our decisions this way. We think that moving, getting married, changing jobs, etc., as decisions, but we don't think about staying in the same place, staying single, keeping the same job etc., as decisions. Or at least we don't think of them as decisions to the same degree.

The third decision bias is the unchangeability bias. The idea here is that when we face large decisions that seem to be immutable (getting married, having kids, moving to a distant place), the permanence of these decisions makes them seem even larger and more frightening. Not to mention that such decisions increase our potential for regret.

With these three biases combined, it's only natural that your son is apprehensive about moving west. Now, the question is, what can you do to help him make this decision? If I were you I would frame the move as "a six-month trial." With this kind of perspective your son would not think of himself as moving (so there is no loss), he would not think of himself as changing his status quo (he would still think of himself as a New Yorker, only temporarily experiencing the West Coast), and the decision to give the West Coast a try would not look so large and daunting. But of course, once he moves to the West Coast, his perspective will shift. Very quickly he will start feeling at home, get used to his new environment, and develop a new status quo. And from that point, any change from the West Coast would look like a large and potentially regretful decision.

Workplace, Experimenting, Happiness

ON THE CURSE
OF KNOWLEDGE

"Relax. You're a famous author—no one expects you
to talk about anything other than yourself."

Dear Dan,

I recently attended a lecture by a well-known academic, and I was amazed and baffled by his inability to communicate even the most basic concepts in his field of expertise. How can such famous experts be so bad at explaining ideas to others? Is this a requirement of academia?

—RACHEL

Here's a game I sometimes play in my class: I ask a few students to think about a song, not to tell anyone what song they picked, and tap the beat of that song on a table. Next I ask the students to predict how many of the students in the room will correctly guess the song's name. They usually think that about half will get it. Then I ask the students who were listening to the beat to name the song that they think was being played, and almost no one gets it right.

The point is that when we know something and know it well (for example, the song that we have picked), it is hard for us to appreciate the gaps in other people's understanding—a bias that is called "the curse of knowledge." We all suffer from this affliction, but it's particularly severe for academics. Why? Because academics study the same topic for years in all its details and intricacies, and by the time we become one of the world's experts on that particular topic, the whole domain seems simpler and more intuitive. And with this curse of knowledge it is easy to assume that everyone else also finds the topic simple and easy to understand.

So maybe the type of difficulties you experienced is indeed something of a professional requirement.

Language, Other People, Predictions

ON BAD SEX

"You haven't a clue which buttons to push."

Dear Dan,

What do you think is worse for a man, if a woman falls asleep during the first time they have sex, or if she starts crying?

—SIIRI

My nonscientific sense is that crying would be much worse. Unless the man could tell himself that the reason for the crying is that she just realized for the first time how wonderful sex can be.

Relationships, Sex, Self-Deception

ON MICE
AND MARKETS

Dear Dan,

Do markets make us more or less moral? On one hand, markets make us think explicitly about other people, which might increase our morality. On the other hand, markets are competitive in nature, which might make people more focused on winning and losing—and less on the fairness of the process. Any insight on this?

—XIMENA

One answer to your question comes from a set of experiments by Armin Falk and Nora Szech. In one of their experiments they asked participants to make a trade-off between saving the life of a mouse that was about to be put to death and earning some money. In the basic condition (the individual condition) participants could either receive no money and save the life of a mouse, or get some money and the mouse would be killed (they were also shown a picture of the mouse and a video about the killing procedure). The decision to save the life of the mouse at a personal expense was compared between this individual condition and two market conditions. In the first market condition (the bilateral market condition) one seller and one buyer negotiated over the killing of the mouse for money. In the second market condition (the multilateral market condition) multiple buyers and sellers negotiated over the killing of the mouse for money.

The results showed that a much larger percentage of participants were willing to kill the mouse in the market conditions (72.2 percent in the bilateral and 75.9 percent in the multilateral) compared with the individual condition (45.9 percent). These results indicate that when we come together in markets, we are more likely to disregard our moral standards for personal gain. Falk and Szech carried out another experiment, one that did not include any moral questions, and they showed that when morality was not involved, there was no difference between individual and market conditions—suggesting that markets directly erode morals, which is certainly not good news for our market-based society.

Stock Market, Morality, Honesty

ON LETTING LOOSE

Dear Dan,

You've talked a lot about how to resist the many temptations that are all around us. I have a similar question, but from the opposite angle.

I'm generally pretty good at making rational decisions that are good for me in the long term and in general I am able to avoid temptation. However, I sometimes take this skill to the extreme. For example, when deciding between whether to watch a movie or make progress on a work-related project, I go for the work-related project, even when I don't feel like it. For some reason, I am not able to bring myself to relax, watch TV, or hang out with friends because I can't help but feel that these activities are a waste of time.

How do you recommend I deal with this need for hyperproductivity?

—DAVE

The feeling you're describing is the need to achieve, make progress, satisfy important general objectives, and be in control. But sometimes we just want to relax and have a good time. Sometimes we need to let go. How can we be who we want to be when the need for achievement and control is so high? I suspect that the most common remedy for this situation is alcohol. Have a fun weekend, and remember to take two aspirin and drink lots of water before going to bed.

Fun, Self-Control, Happiness

ON SHRINKING
AND HONESTY

Dear Dan,

I recently read a study on Italian male sexuality claiming that the average size of Italian male genitalia is roughly 10 percent smaller today than it was fifty years ago. What do you think about this? Is this good or bad news?

—JOHN

The most positive interpretation of these findings is that Italian men have gotten 10 percent more honest in the last fifty years.

Sex, Honesty, Self-Deception

ON HIGH HEELS

Dear Dan,

Why do you think men are attracted to women in high heels?

—ANN-MARIE

There are many possible reasons for this. First, high heels change the posture of those wearing them. Wearing high heels makes people stand straighter, and they push the butt and the chest out a bit. Second, high heels make people look slimmer and taller by changing the width-to-height ratio. Third, they change the shape of the legs: make them look more slender, and the muscles become more defined. I am sure that there are other physical changes as well.

But, personally, I like the evolutionary-type of argument the most—that high heels are appealing to men because at an unconscious level they make men believe that this kind of footwear makes it harder for women to run away from them.

Fashion, Sex, Signaling

ON RULES AS
A WAY TO OVERCOME
NEGATIVE SIGNALS

Dear Dan,

I recently went on a date, and as the evening progressed my date and I were about to become passionate and intimate. I had a condom in my wallet, and I was hoping to use it, but I was worried that informing her about the condom would be taken as an indication that I was counting on having sex on the first date. This seems like a lose-lose situation. Telling her that I had a condom would make her think worse of me, and hiding its existence would mean that we would not have sex. Any advice?

—DAVID

This is indeed a conundrum, and this type of indirect communication is what social scientists call signaling. In general terms, you are fearing that revealing this information will send a negative signal to your romantic partner about your intentions, and maybe even more broadly about you as a person.

The ideal way to overcome this problem would be to eliminate the negative interpretation of this signal altogether. This way, you could inform your date about the availability of a condom without having it count against you as evidence that you are expecting sex. And how might you achieve this? The most direct way of course is to legislate that all young males must carry a condom at all times. With such a rule, your dilemma would be eliminated because you would not be implying anything. You would simply be "obeying the law."

Given that there is no such law currently in place, and that one is unlikely to be enacted anytime soon, you need a more immediate solution. How about if you started an online movement asking young men from all over the world to sign up for a "condom promise" where each man would commit to always carry a condom? This way you could tell your future dates that you are part of this humanitarian movement and that it is out of this social responsibility that you always have a condom with you. In fact, with the help of this benevolent online movement, carrying a condom could be viewed as a positive signal of you being a decent and caring human being.

Decisions, Other People, Signaling

ON TAXES AND MITZVAHS

"I'm combing our finances for all this
disposable income I keep reading we have."

Dear Dan,

I hate tax day. Is there any way to make it more pleasant?

—JAMES

When I became of taxpaying age and started filing my own taxes, all I had to do was to complete the 1040EZ form. For the next few years I loved tax day. It was a day when I got to think about how much money I made, how little I managed to save, how much I gave the government (another way to think about it is to think about how much the government takes, but I prefer the giving framing), and what benefits I got in return from the federal and state governments. It was a good day of financial and civic reflections.

Over the years my taxes have become more complex, more difficult to figure out, and the whole process became more annoying. With these changes, tax day has shifted from a day of reflection on my role and duty as a citizen of this amazing country to a day (or more likely a week) of adversarial relationship with the Internal Revenue Service and the U.S. government.

So, what can we do to make tax day better? The word *mitzvah* in Hebrew means both a duty and a privilege, and one of the things I try to do in the dark hours of struggling with the different forms and rules is to think about taxes as a mitzvah.

Reframing taxes this way is one thing that we, as citizens, can do but I also think that the U.S. government has to do its part. First, the tax code has to become much, much, much simpler if we are to experience tax day as a day of citizenship and not just annoyance. Aside from increased simplicity I think that tax day should be used as an opportunity to educate citizens about where our tax dollars have been spent. Perhaps using some kind of tax receipt where we will be shown the itemized list of our contributions to the operation of the country. Some of the expenses on this list might make us happy, some items will be news to us, and some will make us furious. But it would be a good step toward making tax day a day of civic engagement. A somewhat more extreme version of this idea is to give the citizen the opportunity to vote on where 5 percent of our taxes will go. Let us decide if we want more education, more health care, more infrastructure, etc. Isn't this a step in the direction of what a democracy should be all about?

And let's change the name of this day from tax day to Mitzvah Day.

Spending, Language, Appreciation

ON BULL SERVICE

Dear Dan,

Why do we use the word *service* with things like the Internal Revenue "Service," the U.S. Postal "Service," cable TV "service," and "customer service"?

—YORAM

When I was much younger I got to spend some time on a farm, where I heard farmers saying that they were going to hire a bull to "service" their cows. Maybe this is the answer to your question?

Workplace, Language, Misery

ON LOSS AVERSION
AND SPORTS

Dear Dan,

   You have mentioned many times the principle of loss aversion, where the pain of losing is much higher than the joy of winning. The recent World Cup was most likely the largest spectator event in the history of the world and fans from across the globe were clearly very involved. If indeed, as suggested by loss aversion, people suffer more from losing than they are elated by winning, why would anyone become a fan of a team? After all, as fans we have about an equal chance of losing (which you claim is very painful) and of winning (which you claim does not provide the same extreme emotional impact). So in total, across many games, the outcome for fans is not a good deal. Am I missing something in my application of loss aversion? Is loss aversion not relevant to sports?

—FERNANDO

This description of "fan-ness" implies that people have a choice in the matter, and that they carefully consider the benefits versus the costs of becoming a fan of a particular team. Personally, I suspect that the choice of what team we root for is closer to religious convictions than to rational choice, which means that we don't really make an active choice when choosing a team (at least not a deliberate, informed one) and that we are "given" our team affiliation by our surroundings, family, and friends.

Another assumption implied in your question is that when we approach the choice of a team, we consider the possible negative emotions that would accompany losing relative to the emotional boost of winning. The problem with this part of your argument is that predicting our emotional reactions to losses is something we are not very good at, which means that we are not very likely to accurately take the full effect of loss aversion into account when we make choices.

In your question you also raised the possibility that loss aversion might not apply to sporting events. This is a very interesting possibility, and I would like to speculate why you are (partially) correct. Sporting events are not just about the outcome. If anything, they are more about the ways in which we experience the games as they unfold over time (even the 7–1 Germany versus Brazil game). Unlike monetary gambles, games take some time and the duration of the game itself is arguably what provides the largest part of the enjoyment. To illustrate this idea, consider two individuals: N (Not-Caring) and F (Fan). What loss aversion implies is that N will end up with a neutral feeling regardless of the game's outcome, while F has about an equal chance of being somewhat happy or very upset (and the expected value of these two potential outcomes is negative). But this part of the analysis only takes into account the outcome of the game. What about the enjoyment during the game itself? Here N is not going to get much emotional value during the game. By definition he doesn't care much, and he might spend the time checking his phone or flipping channels. F, on the other hand, is going to experience a lot of ups and downs, feel a connection to the team and to the plays, and will be emotionally engrossed throughout the game. Now, if we take both the experiences of the game and the final outcome into account, we could argue that the serious fans are risking a large and painful disappointment at the end of each game but in doing so they are extracting much more enjoyment from the game itself. And, as in many other areas in life, enjoying the process is often much more important than the final outcome.

Sports, Loss Aversion, Emotions

Acknowledgments

Being trusted by many people with so many questions has been a tremendous privilege on multiple levels. For one, the questions people asked me taught me a lot about what kinds of issues and dilemmas people find puzzling and struggle with. A second source of joy, somewhat like trying to solve a puzzle, came out of analyzing these questions, broadening them to expose the larger principle that they were really about, and then examining what we know and don't know about the topic from the perspective of social science. A third source of satisfaction came from my hope (perhaps naively) that my answers might be interesting or helpful to someone. And finally, I found it very challenging yet satisfying to try to express an idea within a very limited number of words.

The roots of this book is in my *Wall Street Journal* column. And while my editors are never acknowledged publicly in the *Wall Street Journal*, writing this book allows me to express how much they have helped me along the way. Every other week I would get a short lesson on how to write, express ideas more clearly, and of course how to stick to deadlines. My deep unedited thanks goes to my editors: Peter Saenger, Warren Bass, and Gary Rosen.

Because this book involved material that was previously published in the *Wall Street Journal*, it also involved more lawyers than usual. Take the lawyers for the *Wall Street Journal*, add to them the lawyers from HarperCollins, and on top of that add the complexity that I wanted the profits from this book to support research in social science—and you have the conditions that could make anyone lose their mind. Note that I am not thanking any of the lawyers involved, but I do want to express my thanks and admiration to my literary agent, Jim Levine, and his team at Levine Greenberg Rostan for delivering this project against all odds. The team at HarperCollins, led by Claire Wachtel, also deserves praise for their helpfulness, kindness, and patience.

This project would not be as fun to work on, or as interesting to read, without the contribution of William Haefeli. We started this collaboration cautiously, with only one cartoon, testing the idea of working together, but very quickly it became clear that we shared a very similar perspective on many topics, and that we enjoyed the give-and-take between the readers' questions, my answers, and William's insightful take on the topics. I already miss William's emails with suggested cartoons for each week's answers.

My endless gratitude goes to the person who functions as my external memory, hands, and alter ego. The person who always gives me good advice and makes sure that I follow it: Megan Hogerty. My deepest thanks also go to Matt Trower and Aline Grüneisen, who helped organize, sort, and improve the material in this book and made the process collaborative and fun.

Finally, where would I be without my lovely wife, Sumi? Perhaps the best piece of advice that I can give about couplehood is to find someone who you admire and aspire to be like in some important ways—and then spend the rest of your life striving to improve and catch up. Having adopted this advice myself, I can personally attest to how magical it can make one's life.

I spend much of my time traveling around the world, talking about my research, and giving advice to which people seem to be listening. With this kind of lifestyle, there is a real risk of developing an inflated ego. But then I get back home and very quickly I am reminded of how much more I need to learn about true intelligence, kindness, and generosity.

Loving, Dan

Categories

About the Author

Dan Ariely is the James B. Duke Professor of Psychology and Behavioral Economics at Duke University, and is the founder of the Center for Advanced Hindsight. He is the author of three *New York Times* bestsellers: *Predictably Irrational*, *The Upside of Irrationality*, and *The Honest Truth About Dishonesty*. He lives in Durham, North Carolina, with his wife, Sumi, and their two adorable and well-behaved children, Amit and Neta.

William Haefeli is an internationally revered cartoonist. He is a cartoonist for *The New Yorker*, and his work has also appeared in numerous magazines in the United States and abroad, including *Punch*, *The Advocate*, *The London Magazine*, *Chicago*, and *Saturday Review*. William studied psychology at Duke University and art at the Chicago Academy of Fine Arts.

BOOKS BY DAN ARIELY

THE (HONEST) TRUTH ABOUT DISHONESTY
How We Lie to Everyone—Especially Ourselves

Available in Paperback and eBook

"A lively tour through the impulses that cause many of us to cheat, the book offers especially keen insights into the ways in which we cut corners while still thinking of ourselves as moral people."
—*Time*

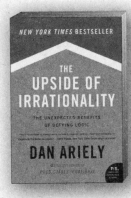

THE UPSIDE OF IRRATIONALITY
The Unexpected Benefits of Defying Logic

Available in Paperback, eBook, and Audio CD

"*The Upside of Irrationality* is an eye-opening, insightful look at human behavior, proving that defying logic is part of what makes us human."
—*Boston Globe*

"Deciding how to apply [Ariely's] insights is a pleasure that lingers long after the book is finished."
—*New York Times Book Review*

PREDICTABLY IRRATIONAL
The Hidden Forces That Shape Our Decisions

Available in Paperback, eBook, and Audio CD

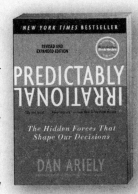

"A fascinating romp through the science of decision-making that unmasks the ways that emotions, social norms, expectations, and context lead us astray." —*Time*

"Surprisingly entertaining . . . easy to read . . . Ariely's book makes economics and the strange happenings of the human mind fun."
—*USA Today*